KT-598-826

CONSERVATION 2000

The Acid Rain Effect

First published 1992

Typeset by J&L Composition Ltd, Filey, North Yorkshire
and printed in Great Britain by
Bath Press Colourbooks, Blantyre

for the publishers
B.T. Batsford Ltd
4 Fitzhardinge Street
London W1H 0AH
ISBN 0 7134 6501 8

A CIP catalogue record for this book is available from the British Library

Acknowledgements

The author would like to thank the following for their help in the preparation of this book: Acid Rain Information Centre, Manchester; Christer Agren; Canadian Embassy; Dr Alastair Donald, National Rivers Authority, Cardiff; Lauren Gillen, US Environmental Protection Agency; F. John Smith, Johnson Matthey plc.
The author and publishers would also like to thank the following for permission to reproduce illustrations: Canadian Embassy page 9; CEGB pages 38 & 52; Green & Sons Ltd page 46; Sarah Harding page 37; Johnson Matthey plc pages 54–55; London Environmental Bulletin page 12; National Rivers Authority, Llyn Brianne Project pages 34 & 50; National Swedish Environment Protection Board page 13; Joy Palmer page 58; Sheffield City Library page 21; Swedish NGO Secretariat on Acid Rain page 60; US Department of the Interior page 33; US Environmental Protection Agency page 44; Warren Springs Laboratory, Department of the Environment pages 8, 15, 19, 23 & 27; Line drawings by Ken Smith.

CONSERVATION 2000

The Acid Rain Effect

Philip Neal

B.T. Batsford Ltd, London

CONTENTS

Note: the formulation SO_2 for sulphur dioxide and NOx for nitrogen oxides are used throughout this book.

INTRODUCTION: WHAT IS ACID RAIN?

When something is incinerated (burnt), heat is created, gases and unburnt particles rise into the air and heavier waste falls to the ground as ash. Material is burnt for warmth or light or so that it can be destroyed – only ash and heavy flame-resistant matter is left: the smoke and gas have risen into the atmosphere; it is these fumes that cause ACID RAIN.

Acid Rain is 'ordinary rain' with a high acidity which acts like a weak acid. We shall see what this means later. But snow, hail and sleet also fall from the sky, and mist or fog float like clouds at ground level. All of these can be acidic as well – so we could call it ACID PRECIPITATION instead. Yet some of the gases and some of the unburnt particles fall back to earth very quickly, before they have a chance to react with water in the atmosphere: they are dry when they touch down. When rain arrives they become weak acids and have a damaging effect on whatever they touch. It would be better, therefore, if we referred to all of these types of acidic moisture not just as Acid Rain but as ACID DEPOSITION. This does not quite have the urgent popular message of the term ACID RAIN, so perhaps in order to make as many people aware of the problem as possible it may be best to continue to deal with this serious environmental problem as Acid Rain, though it is important to be aware that it is not just rain that is involved.

This book will try to help you understand the problem, and how it can be prevented. Although most of the preventive measures involve the improvement of industrial processes and vehicle technology, there are some things that individuals can do to help reduce Acid Rain. New regulations, new scientific discoveries, new ways of producing power – all of which may influence the production of Acid Rain – are occurring regularly. Many newspapers have environmental sections nowadays, and there are also an increasing number of environmental magazines. Reading these are some of the ways to keep up with developments concerning the problem of Acid Rain and other environmental issues.

HOW DOES ACID RAIN WORK?

TRANSPORTATION

EMISSION

Emission

Most of the power for industry, transport and our homes comes from burning coal or oil. Each can be used directly as a fuel or used to generate electricity. Not all of the fuel is burnt. Some of it becomes ash or waste oil and much of it is sent into the air as unburnt particles or as gas. These exhaust fumes come from the engines of cars, trucks and other vehicles; from aeroplanes, trains and boats; from the chimneys (smoke stacks) of power stations, factories and houses. Smoke clouds also rise from volcanoes, forest fires and other natural sources. All of these fumes are called EMISSIONS.

TRANSFORMATION

DEPOSITION

Transportation

The emissions rise into the air, the heavier unburnt particles and some of the gases fall back quickly to earth – near the place from which they rose. The lighter particles and most of the gases are caught by the wind and carried over land and sea, from one town to another, and from one country to another. There are no fences in the sky and no customs barriers. The fumes are carried up through the clouds into the upper air and on for mile after mile. TRANSPORTATION has taken place.

Transformation

As the emissions rise through the clouds and fall back to earth through other clouds, chemical reactions occur. The *aerosols* (drops) of water floating in the air (which make up the clouds) mix with the gases and are changed into weak acids. The chemical processes are many and varied and result in an increase in the acidity of the water droplets. TRANSFORMATION is the term given to this change.

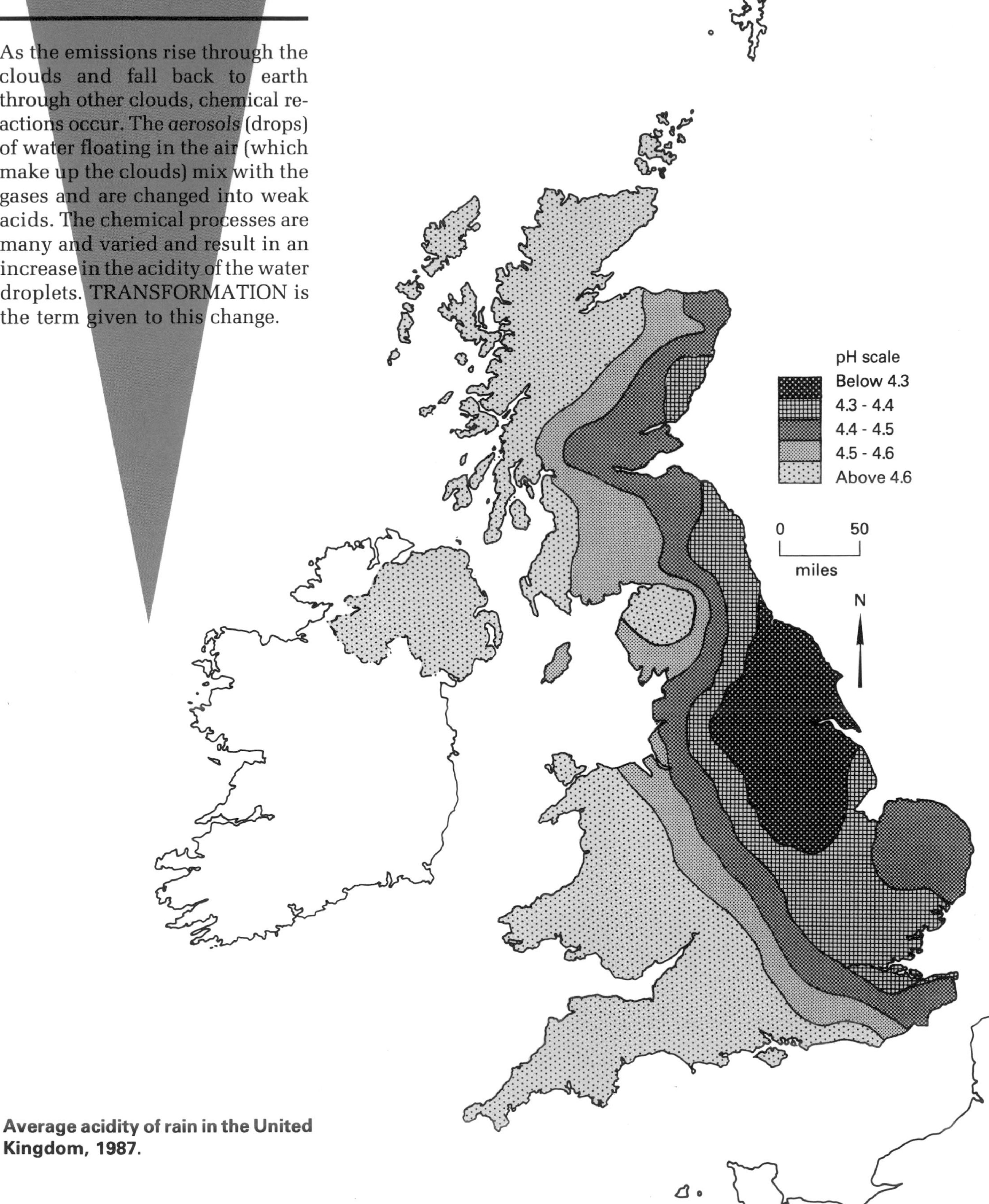

Average acidity of rain in the United Kingdom, 1987.

Deposition

As the droplets become larger through collision one with the other, following the normal water cycle of nature, so they begin to fall to earth as precipitation – rain, hail, sleet or snow. Eventually they reach the ground to fall on buildings, trees, roads, grass, oceans or whatever is in the way. The acidic drops shower down on to the earth. This is called DEPOSITION of Acid Rain. Those portions of the emissions that fall directly to the ground without reaching the cloud layer are dry when they touch down: this is DRY DEPOSITION. When the dry deposit mixes with water it causes extra acidity.

Who pays?

One of the problems of Acid Rain pollution is that the country which causes it does not necessarily suffer the consequences. The guilty emitters may well be reluctant to pay to prevent Acid Rain. Why spend hard-earned dollars, marks, pounds, yen, francs, kroner, lire – or whatever the currency may be – when the harm is done somewhere else?

Guilty gases

The damaging gases of Acid Rain are mainly sulphur dioxide (SO_2) and nitrogen oxides (NOx). Reacting chemically with water (H_2O), they become sulphuric acid and nitric and nitrous acids. The natural emissions contain carbon gas, carbon dioxide (CO_2) and carbon monoxide (CO), as do the gases from the burning of fuel. From these comes carbonic acid, which is sufficiently present in the atmosphere for most precipitation to be naturally acidic to some extent.

What is acidity? We shall look at this separately but it must be emphasized that the acids of Acid Rain are very weak. Sulphuric acid when concentrated is a powerful destroyer. It burns away flesh and wood and anything organic. The very dilute sulphuric acid of Acid Rain never does this – but even at low strength it can weaken many things and bring about their destruction over time.

Acid Rain in the USA and Canada measured in the amount of sulphur deposited in kilograms per hectare of land. 20 kg per hectare is about pH 4.5. Average readings 1984–7.

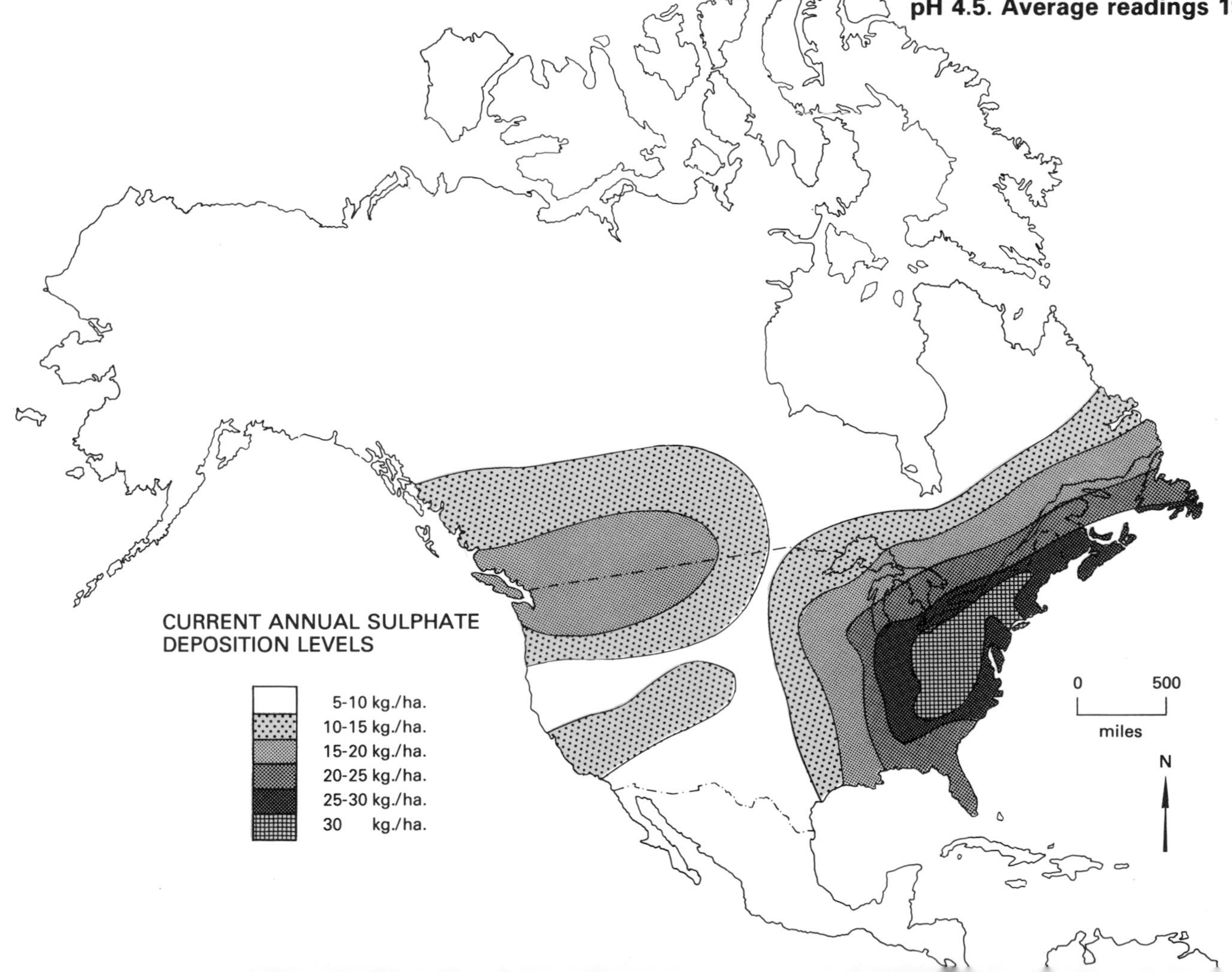

HOW IS ACIDITY MEASURED?

How acid a substance is can be measured by using the pH scale. This indicates the concentration of hydrogen ions present in the test solution and is measured on a *reciprocal logarithmic scale*. All this means, for the purpose of the investigator, is that each unit on the scale represents a tenfold difference in acidity. In other words an increase from 5 to 6 on the scale tells you that the solution is 10 times more concentrated and between 5 and 7 it is 100 times more concentrated.

The pH scale goes from its *most acidic* at the *low number end* to its *most non-acidic* (or *basic* or *alkaline*) at the *high number end*. The actual scale is between 0 and 14. pH 7 is neutral, the measure of distilled water – water with all of its impurities removed. Everyday substances have an acidity and some are shown in the diagram.

It follows from this that very high acidity on the pH scale is not necessarily harmful to our bodies. Some people douse their chips in vinegar and squeeze lemon juice on to their fish. The liquid that is swallowed (and which does no particular harm) has a pH measure of about 2.5. If this was a

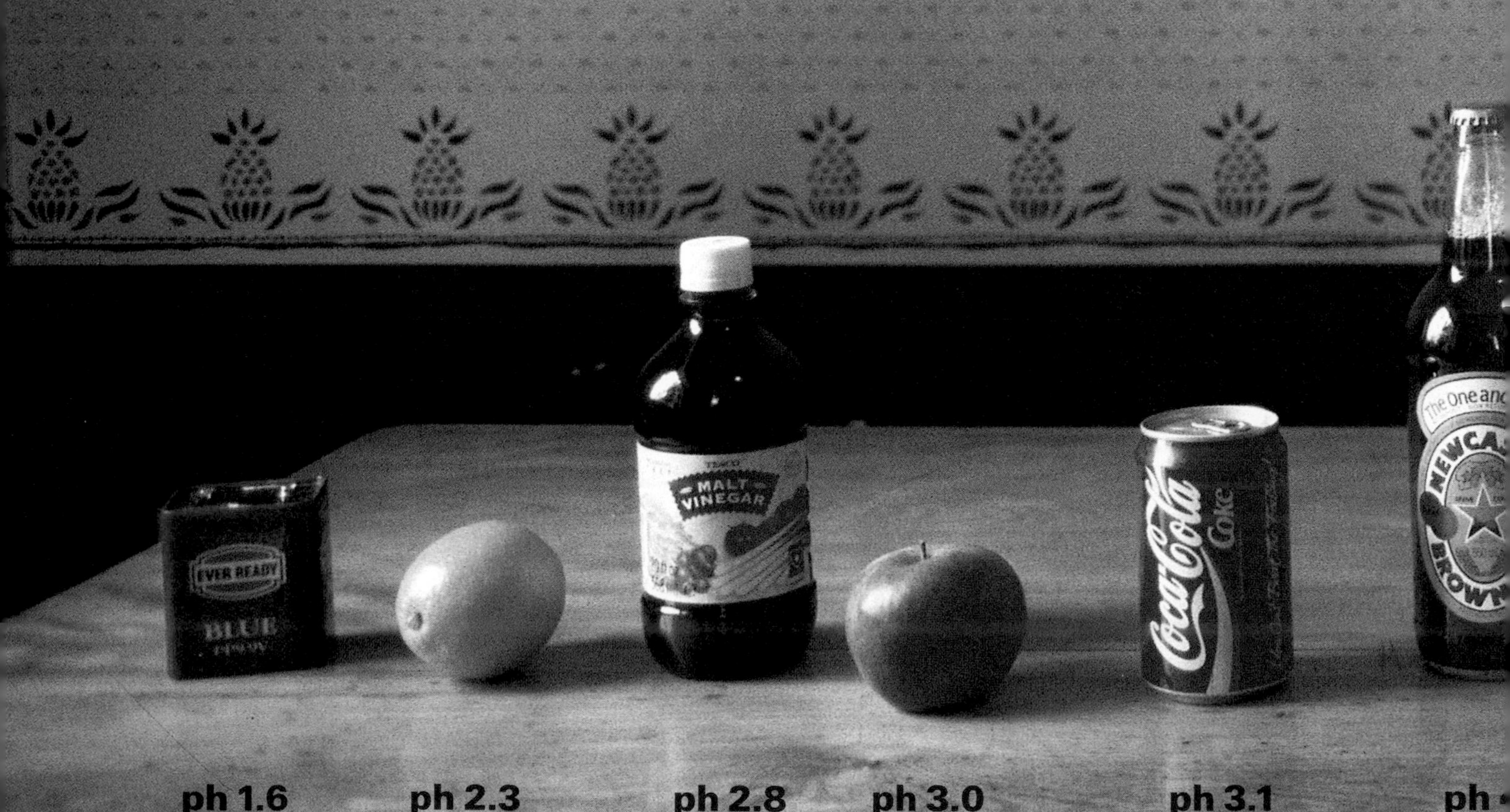

measurement of rainfall it would be said to be highly acidic – but it is still far from our ideas of acids burning holes in the carpet. Acid Rain will not make holes in your umbrella – but it can do much harm elsewhere, though as a liquid it is still only a very weak acid.

Usually rainfall in the UK or USA has a pH reading of between 5 and 6, averaging out at 5.6. This is due to the fact that carbon gases occur commonly in the air, so that rain falling through it reacts to become a weak carbonic acid.

pH

pH gets its name from the Danish scientist Sorenson who first suggested the scale. He used the word 'potenz', the Danish word for power, and H for hydrogen, which he abbreviated to 'pH'. If you think of 'p' to stand for 'proportion' this will remind you of its real meaning as a measurement.

Testing river water for pollutants, including acid levels.

Measuring acidity

pH is most simply measured by using *indicators*, which are substances which change colour when in contact with acid or alkaline materials. Litmus is a good example. This is a purple dye obtained from certain lichens, which changes colour depending on the acidity of the liquid in which it is dissolved. Litmus paper is impregnated with this dye and changes colour (red for acid and blue for alkaline) when in contact with liquids of various states of acidity. *Universal indicators* are another type of dye: papers made from them are used to give an actual pH value. The different colour which results when a paper is dipped into the test liquid is then compared with a colour chart and the pH value read off. Farmers and gardeners are very interested in the acidity of their soils, and they use pH indicator kits to investigate it. Simple kits can be bought at most garden centres. Usually soil is mixed with distilled water and the universal test indicator is added, either as a liquid or a paper strip.

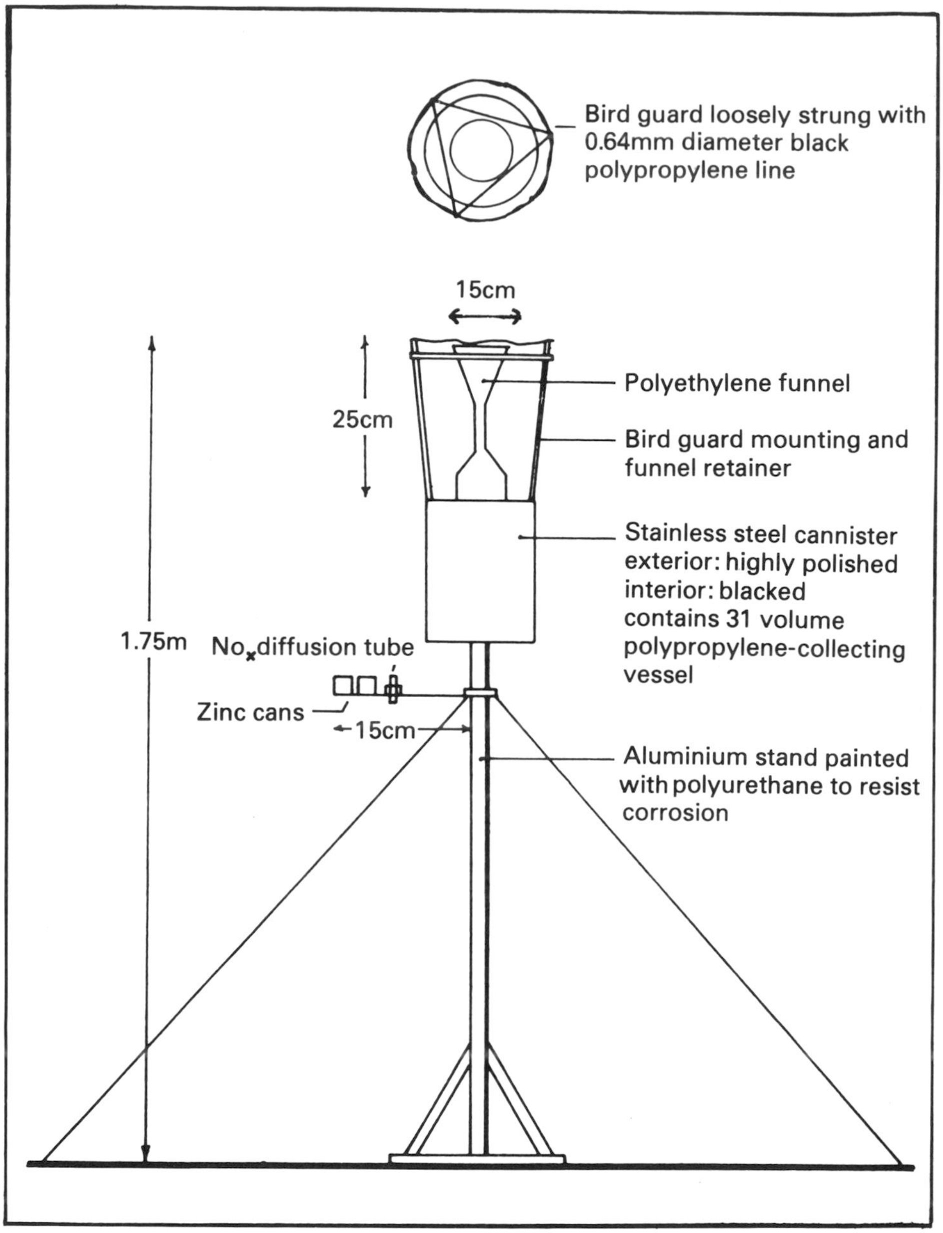

Bird guard loosely strung with 0.64mm diameter black polypropylene line
15cm
25cm
Polyethylene funnel
Bird guard mounting and funnel retainer
Stainless steel cannister exterior: highly polished interior: blacked contains 31 volume polypropylene-collecting vessel
1.75m
NO_x diffusion tube
Zinc cans
15cm
Aluminium stand painted with polyurethane to resist corrosion

Acid Rain collector designed to avoid any acidity being 'picked up' from surrounds. Note the NOx tube which measures the amounts of the gas in the air and the zinc cans which will indicate acid damage to their surface.

Carrying out an acidity test for rainfall

Remember that it is the acidity of the rain which you want to record and not that of the container in which it is collected. It is very important, therefore, that jars or buckets are clean. To ensure this it is necessary to wash the container thoroughly with distilled water and to place the collector in such a position that other sources of acidity do not contaminate the sample. Place the collector well away from trees, roofs, overhead cables and hedges so that the rain which gets into the container has not dripped down from something above. Put it above the ground so that splashes do not affect the collected water. It is no use taking the water sample from a butt which collects the flow from a gutter, as it will be affected by chemicals from the roofing material, the gutter and drainpipes.

Remember, too, that rainfall will be most acidic at the start of a storm, as the air will be loaded with the collected pollution of the days since it last rained. It will be best to make several collections during the time that it is raining and to take the average of the readings as the acid level of the rain. However it is important to measure the first rain that falls after a long period of dry weather, as its high acidity will cause more damage than later rain.

Carrying out an acidity test for a lake or river

The water in a lake or in a river can vary in acidity at different places. It is necessary to collect samples from different locations – from the middle as well as from close to the bank. Collect samples from farther out by using a collecting jar on a long pole, rather than by wading into the water. Samples can also be collected from a boat under adult supervision. I assume that anybody collecting such samples will be able to swim – if not, learn to do that before worrying about acidic ponds! Collect samples over a period of several days in order to obtain a measure of average acidity. This will mean that any 'odd' readings will be eliminated and so give you a more accurate acidity result.

Area of lake:
9,500 hectares (36.7 square miles)

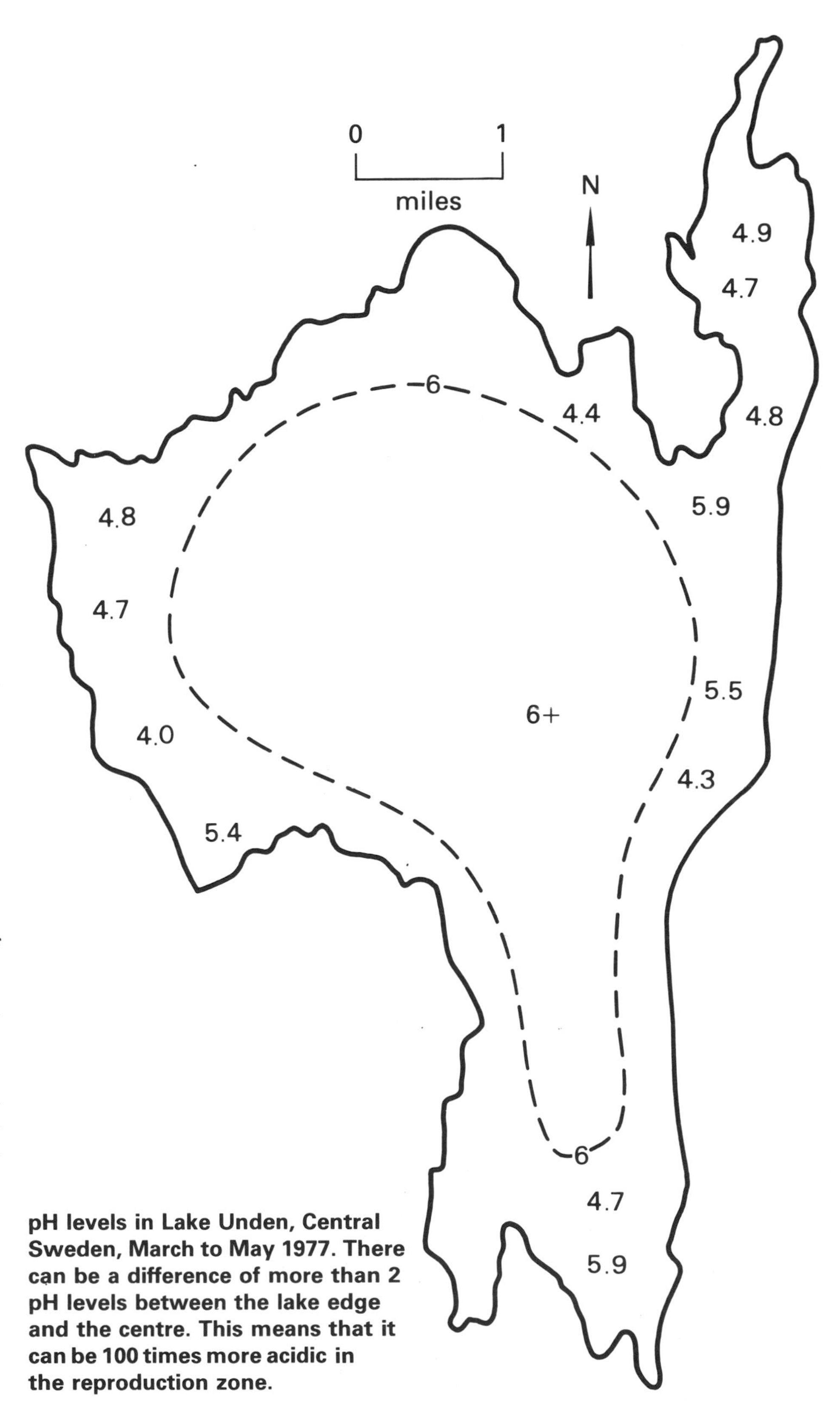

pH levels in Lake Unden, Central Sweden, March to May 1977. There can be a difference of more than 2 pH levels between the lake edge and the centre. This means that it can be 100 times more acidic in the reproduction zone.

CAUSES: POWER STATIONS AND NATURAL SOURCES

Over 35 coal-burning power stations are in operation in the United Kingdom and several hundred in the USA. Each one has a chimney – a very tall chimney in the most modern. From these rise the unwanted gases and other waste which has not been trapped in the filtering systems. Minute after minute, week after week, year after year the fumes pour forth, sometimes clearly visible, sometimes less obvious but present all the same.

As I travel to work I see, every day – not only once or twice, but every day – a plume of smoke surrounded by clouds of steam; the smoke from a chimney and the steam from the cooling towers of a power station. This is one of the places where electricity is made – and one of the places where Acid Rain is born. To put it very simply – somewhere, someone or something is going to suffer damage from Acid Rain so that people, for example, can watch television programmes or use their washing machines. We must not ignore the natural sources of SO_2, NOx and CO_2 which are present in the atmosphere.

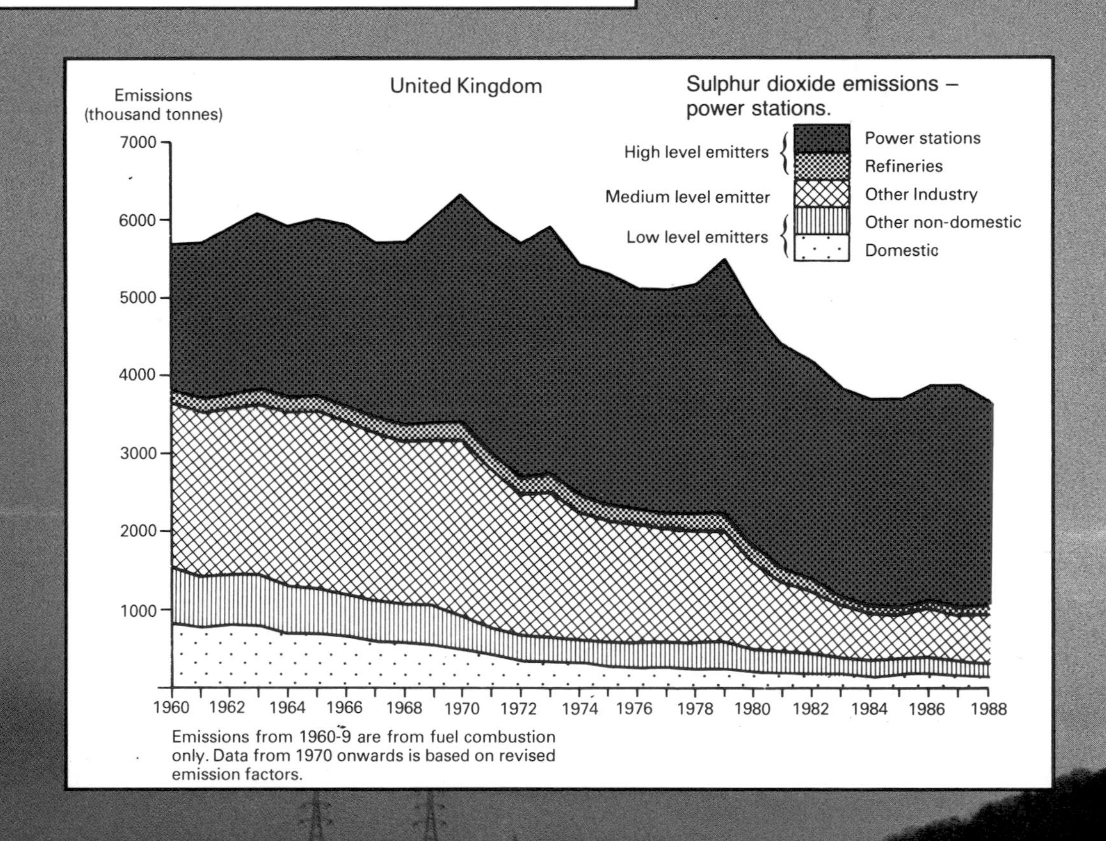

Emissions from 1960-9 are from fuel combustion only. Data from 1970 onwards is based on revised emission factors.

Power stations

It would not be so bad if power stations were efficient. For every three tonnes of coal burnt about two tonnes are wasted in the process of generating electricity, most of it in the heat lost in the cooling system. What a waste: it does not have to be; *Combined Heat and Power* means just that, making electric power but at the same time using the heat that would be wasted for other purposes for example, to heat nearby houses and provide their occupants with hot water. They can do it in Scandinavian countries – why can't others?

But if it is bad in Great Britain and some parts of the USA, it is even worse in other parts of Europe. Although much of the coal burnt in the UK and the USA contains more sulphur than some coals which are mined elsewhere, at least those countries do not use a lot of *Lignite*, often known as *Brown Coal*. The eastern part of Germany, Czechoslovakia and other East European countries, use it extensively. Usually it is high in sulphur, so that with over a third of the electric power in Czechoslovakia coming from lignite burning, it is no wonder that not only do pollutant gases create Acid Rain which has devastated much of the forest in that country, but that the recent revolution against the Communist government may have been as much about pollution as about politics. In Leipzig, in the former East Germany, there are five power stations burning some of the 314 million tonnes of brown coal mined every year in

that country. At only one of these is there any attempt to install anti-pollution equipment.

Oil-fired power stations tend to produce smaller quantities of the acid-causing gases. Natural gas is an alternative fuel for power stations: gas contributes practically no sulphur emissions, no dust particles and it produces only half as much carbon dioxide as coal for the same amount of energy. The total NOx emissions from gas usage are small compared with coal (see page 49).

Natural sources of SO_2 and NOx

At a rough approximation about half of the SO_2 and NOx in the atmosphere comes from natural sources and the other half from sources created by people. This is not spread around the world evenly. Natural sources are less important where most industry and towns occur. It is estimated that about 90 per cent of the SO_2 in the air in Europe is from non-natural sources and that something like 75 per cent of NOx is also created by human activities. It is similar in the North East USA. Although natural resources contribute to the acidity of rain it is the enormous extra amount of pollutant gases from man-made actions which make for high acidity in these areas.

Algae

Some marine algae can produce dimethylsulphoxide (DMS). Measurements have been made in the North Sea to show that the sulphurous gas which rises from these algae in spring and summer is equal to a quarter of the man-made sulphur in the atmosphere. The westerly winds take this sulphur over Norway and Sweden and it cannot be distinguished from the sulphur from the power stations of Britain. Perhaps Acid Rain in Scandinavia is not all the fault of chimney emissions after all!

A worldwide problem

'The constant sore throats and headaches, the high incidence of lung disease and cancer, the lack of red blood corpuscles in the children, the low life expectancy were a nagging irritant – a daily personal complaint against the old regime.' *The Observer*, 28 January 1990 commenting on air pollution in Czechoslovakia.

CAUSES: INDUSTRY

The manufacture of raw materials which are used by other industries often involves heat. Clay, for example, is pressed into various shapes and then 'cooked' to very high temperatures in kilns to make the hard tiles and bricks for building work. Cement is made by mixing chalk, limestone and coal to be burnt until 'clinker' forms, which is then crushed into cement powder. Coal itself is heated to make coke, coal gas or smokeless fuel. Iron ore has to be melted by great heat to obtain the metal and more heat is required to change it into steel. Bauxite into aluminium, copper ore into copper, tin ore into tin – all require to be melted by the use of furnaces.

Nowadays a lot of heating is by gas, oil or electricity, but burning coal is still the most important source of heat. All of these industrial processes have factories with chimneys to dispose of the exhaust emissions. Modern cleaning techniques have meant that fewer manufacturers need to have smoke pouring from the chimneys – but the less visible gases are still there.

Industries which do not use coal as a fuel still produce gases from the processes of manufacture. The chemical industry – and this includes oil refineries – have factories with a whole array of chimneys and towers of different shapes and sizes. Fumes rise into the air from all of them.

United Kingdom
Sulphur dioxide emissions – industry.
Emissions (thousand tonnes)
7000
6000
5000
4000
3000
2000
1000
1960 1962 1964 1966 1968 1970 1972 1974 1976 1978 1980 1982 1984 1986 1988
High level emitters
Power stations
Refineries
Medium level emitter
Other Industry
Low level emitters
Other non-domestic
Domestic
Emissions from 1960-9 are from fuel combustion only. Data from 1970 onwards is based on revised emission factors.

Removing the particulates

There are few chimneys now in Western European countries, the USA, Canada or Japan which are spewing out untreated smoke. Most of the unburnt pieces, the *particulates*, have been removed: these are the particles which, together with some of the gas, fall on the local area as Dry Deposition.

In some factories the particulates are removed by passing the smoke through water showers where the smoke is washed by a constant stream of water. Unfortunately this method causes a stream of 'muddy' water that has, in turn, to be dealt with. Another way is to use the power of *electrostatic electricity* to attract dust. This is the type of electricity which causes nylon material to spark after it has been worn or dust to settle on the face of the television screen when it has been wiped over with a dry duster. It is easy to make electrostatic electricity by rubbing a plastic comb or pen up and down your sleeve: afterwards pass it over some small pieces of paper and watch them jump on to the object, just like iron filings jumping on to a magnet. In a similar way particulates stick to electrostatically charged screens hanging in the pathway of smoke – a good shake and the dry bits and pieces fall to the bottom – but the sulphur, nitrogen and carbon gases remain.

Much is done by industry in some parts of the world to clean the emissions from factory chimneys but this is mainly confined to the richer countries of North America, Europe, Japan and Australasia. A vast number of coal burning manufacturers still pour untreated smoke and fumes into the air – emissions which spread far beyond their own borders.

Attercliffe steelworks, Sheffield about 100 years ago. Note the many smokestacks and their lack of height.

Tall stacks

Perhaps you have noticed that really old factories have several chimneys about twice the height of the buildings alongside. Others which were built over fifty years ago have rather taller chimneys. The latest factories, especially power stations, have a single chimney which is very tall. This *Tall Stack Policy* (TSP) was a deliberate move to deal with local pollution. The extra height means that the emissions are carried into the upper atmosphere. In the USA and Canada there were two chimneys over 150 metres high in 1970. By 1980 there were over 200: the TSP had been applied.

Careful observation will often show that extra height has been added to an old chimney. This is revealed by newer bricks at the top of a stack or even by metal extensions. In some cases there are two or more additions.

This policy may have remedied some of the local dry deposition but it created much of the problem of the wide dispersal of acidic pollution to places without smoking chimneys of their own.

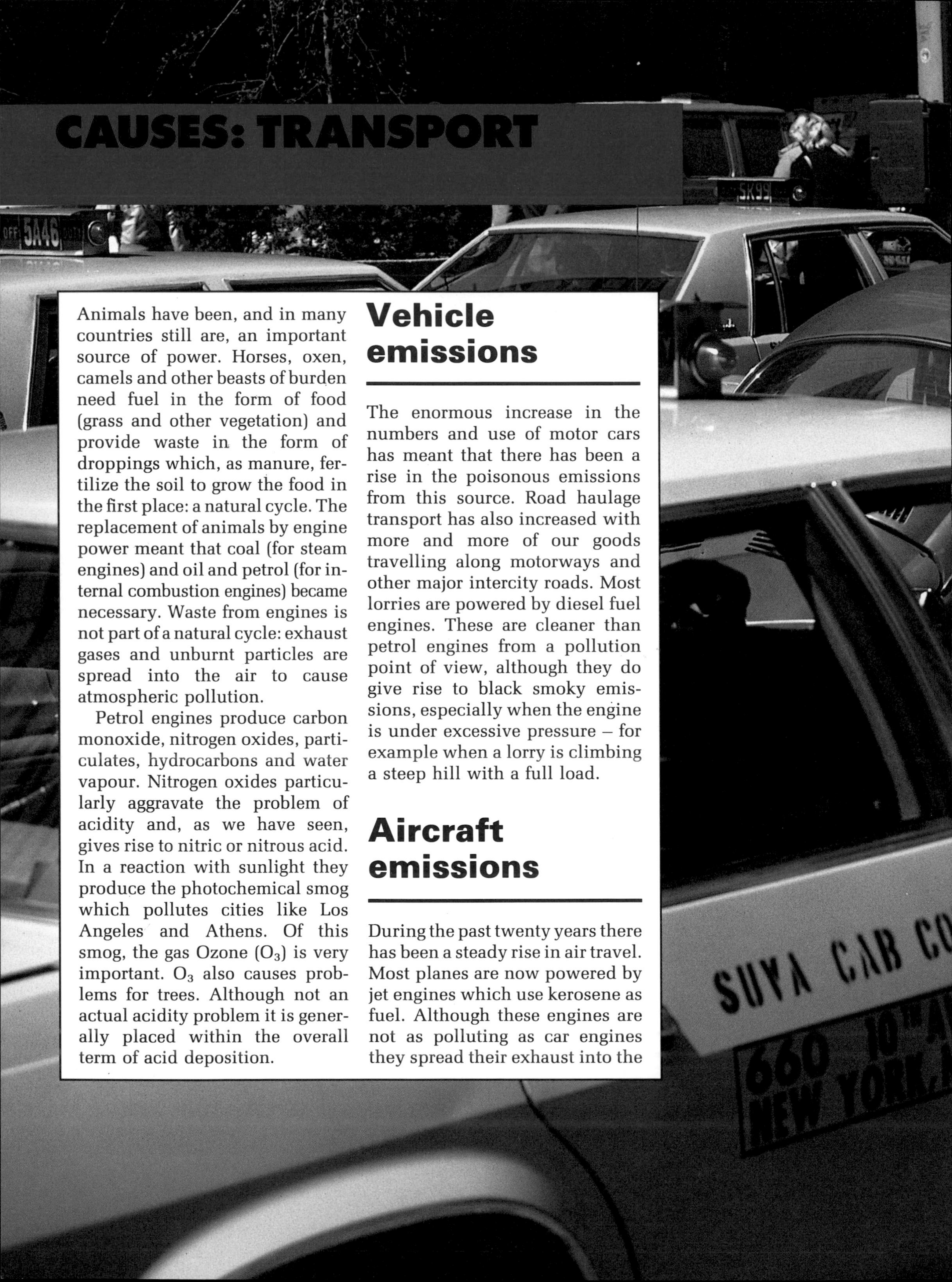

CAUSES: TRANSPORT

Animals have been, and in many countries still are, an important source of power. Horses, oxen, camels and other beasts of burden need fuel in the form of food (grass and other vegetation) and provide waste in the form of droppings which, as manure, fertilize the soil to grow the food in the first place: a natural cycle. The replacement of animals by engine power meant that coal (for steam engines) and oil and petrol (for internal combustion engines) became necessary. Waste from engines is not part of a natural cycle: exhaust gases and unburnt particles are spread into the air to cause atmospheric pollution.

Petrol engines produce carbon monoxide, nitrogen oxides, particulates, hydrocarbons and water vapour. Nitrogen oxides particularly aggravate the problem of acidity and, as we have seen, gives rise to nitric or nitrous acid. In a reaction with sunlight they produce the photochemical smog which pollutes cities like Los Angeles and Athens. Of this smog, the gas Ozone (O_3) is very important. O_3 also causes problems for trees. Although not an actual acidity problem it is generally placed within the overall term of acid deposition.

Vehicle emissions

The enormous increase in the numbers and use of motor cars has meant that there has been a rise in the poisonous emissions from this source. Road haulage transport has also increased with more and more of our goods travelling along motorways and other major intercity roads. Most lorries are powered by diesel fuel engines. These are cleaner than petrol engines from a pollution point of view, although they do give rise to black smoky emissions, especially when the engine is under excessive pressure – for example when a lorry is climbing a steep hill with a full load.

Aircraft emissions

During the past twenty years there has been a steady rise in air travel. Most planes are now powered by jet engines which use kerosene as fuel. Although these engines are not as polluting as car engines they spread their exhaust into the air well above the ground, and spread them in a trail behind them as they move from one place to another. Supersonic aircraft, such as Concorde and military planes, use the upper air so that their emissions are likely to travel around the whole planet.

The increase in transport by road and air means an increase in the amount of oil which has to be refined to make the petrol and the kerosene. As we have seen already, this leads to gas emissions from the refineries.

Vehicle engines

Engines which are idling – that is, running at a low number of revolutions per minute (rpm) – emit the most pollutants. Engines which are turning at about 2000 to 4000 rpm are most efficient and

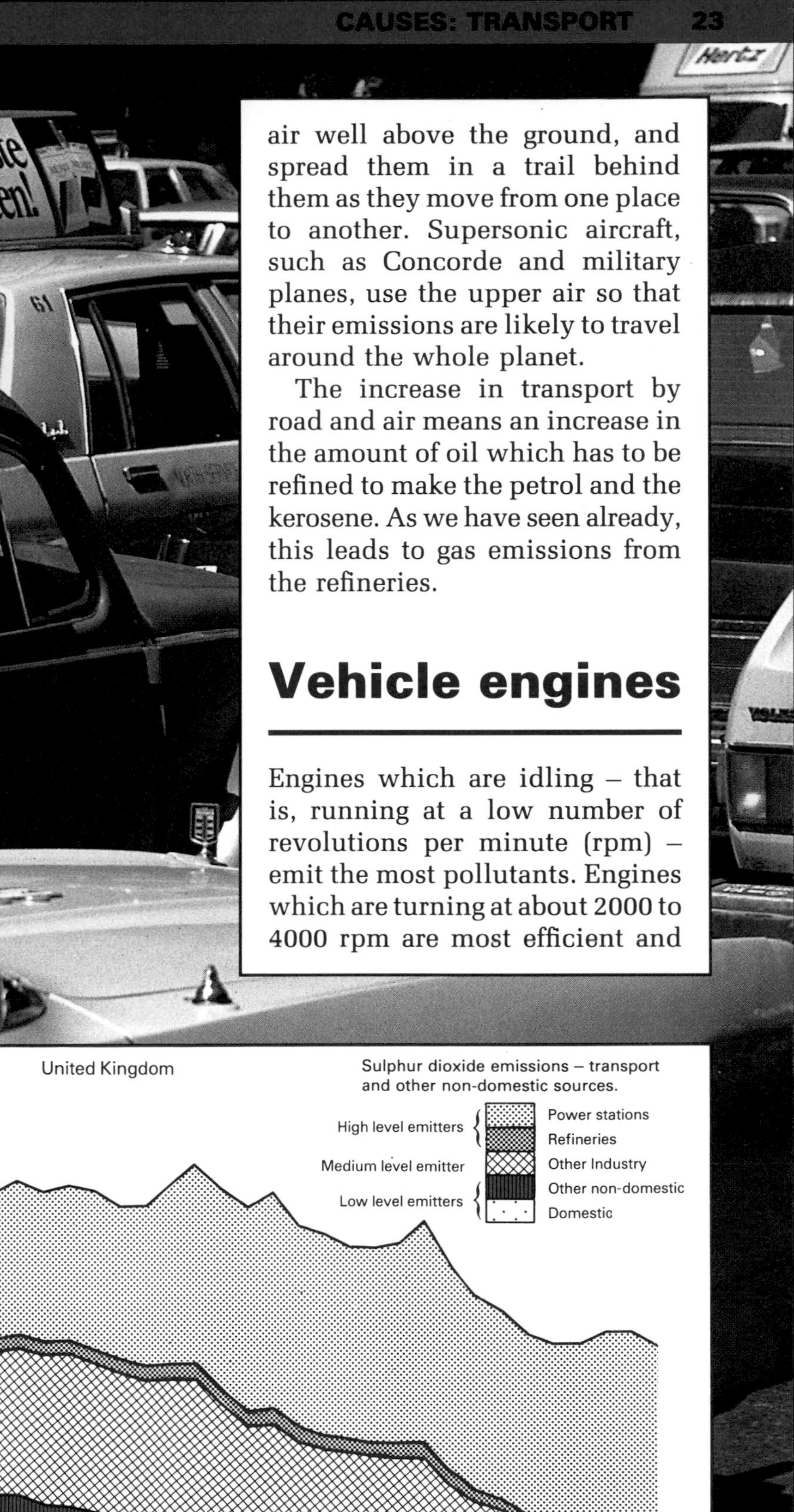

Emissions from 1960-9 are from fuel combustion only. Data from 1970 onwards is based on revised emission factors.

are sending out the least amount of exhaust. A vehicle whose engine is turning at this rate will travel at between 30 and 60 miles per hour (mph) – both fewer revolutions or more revolutions cause greater amounts of exhaust. If you know about driving a vehicle you will know that a lower gear causes the engine to turn over more quickly for less road speed and you will also know that, in a higher gear, as you travel faster the engine has to turn more quickly.

All this means that most pollutant gases are given out when the vehicle:

1 is still or slowly moving as in a traffic jam
2 is travelling in a low gear
3 is travelling very fast
4 is starting up
5 is picking up speed

Driving techniques

Thus it follows that the way we drive affects the amount of pollution: we should aim to move at a steady speed, and to anticipate what might cause us to stop ahead so that we can slow, avoid the halt and then pick up speed again gradually. If we do travel at a high speed on a motorway we should be in as high a gear as possible which makes cars with a 5th gear less polluting than one where the top gear is 4th. AND we need to avoid traffic jams – this is not easy but travelling out of peak times helps. Does everybody need to go to the same leisure attraction at the same time on the same day?

All of this helps to reduce Acid Rain – and it saves money on petrol or diesel as well as time and temper. Do you think that speed limits should be reduced, especially on motorways?

Ozone

Ordinary oxygen gas is made up of two individual atoms of oxygen (O_2): *Ozone* is made up of three atoms (O_3). Ozone in the upper atmosphere acts as a filter to ultraviolet (UV) rays, particularly the UV B variety which can cause skin damage and skin cancer in people with fair skins. At ground level ozone is harmful. It aggravates the eyes, nose and throat, leads to chest discomfort, headaches and coughs. Bronchitis and asthma sufferers can be affected badly.

Ozone is brought about by the reaction of sunlight on pollutant gases especially nitrogen oxides from motor vehicles. If the weather conditions are such that the air is still and hot with little cloud – the sort caused by a high-pressure system – the ozone will collect as part of photochemical smog. Towns, with their mass of road traffic, have the greatest problems. London, Birmingham, New York, Paris and Rome all suffer from the problem when conditions

Smoke and exhaust fumes go upwards because they are warmer than the air around them. Warm air rises. Sometimes there is a warmer layer of air above the ground which traps the fumes. This is called a temperature inversion and the warmer air an inversion layer. These diagrams could represent Los Angeles.

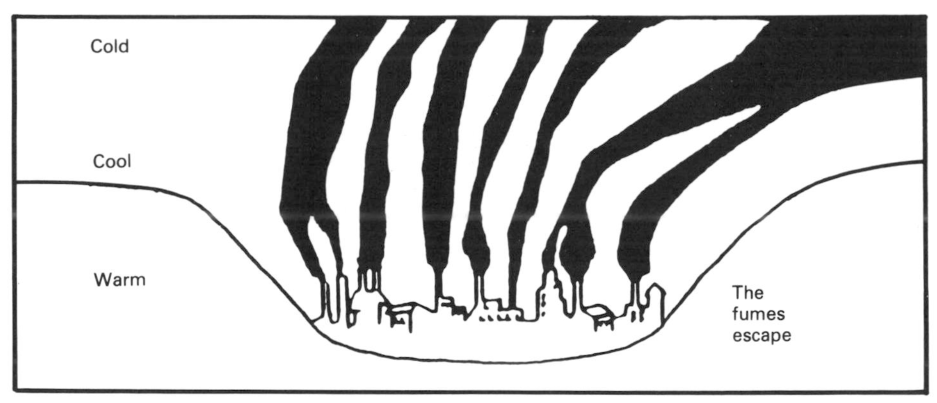

Normal conditions

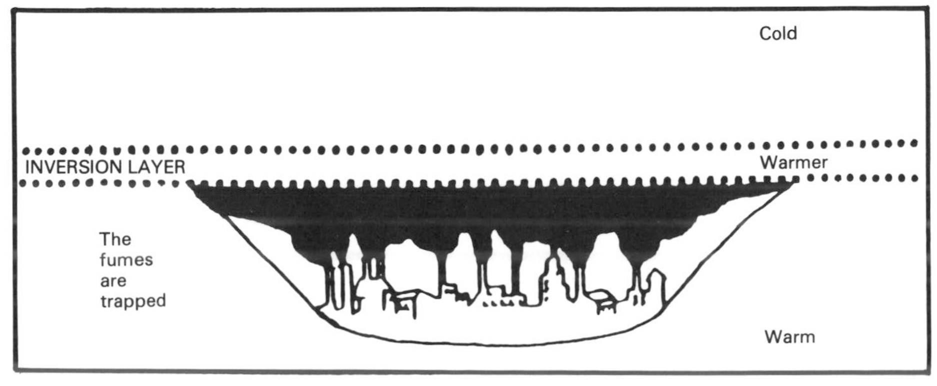

Inversion conditions

are 'right'. If the urban area is situated in a basin surrounded with hills then it is even worse – Los Angeles and Athens are two of the greatest sufferers. The World Health Organization (WHO) recommends a maximum concentration of 50 to 60 parts per thousand million (pptm) of ozone to air as their air quality guideline. In London during a heatwave of May 1990 the level reached 117 pptm: even on the Cornish coast, over two hundred miles from London, the level reached 73 pptm with 7 hours of over 60 pptm on 1 May. As well as weather forecasts and pollen count warnings some daily papers in the United Kingdom are now issuing 'ozone levels' where an incident has occurred or when one is forecast – this enables people at risk to take the necessary precautions. Newspapers in the United States have done this for many years. Cyclists and pedestrians in many urban areas now wear smog masks when conditions are very bad.

Steam trains are mostly things of the past. Diesel locomotives emit fumes: electric trains are pollution-free.

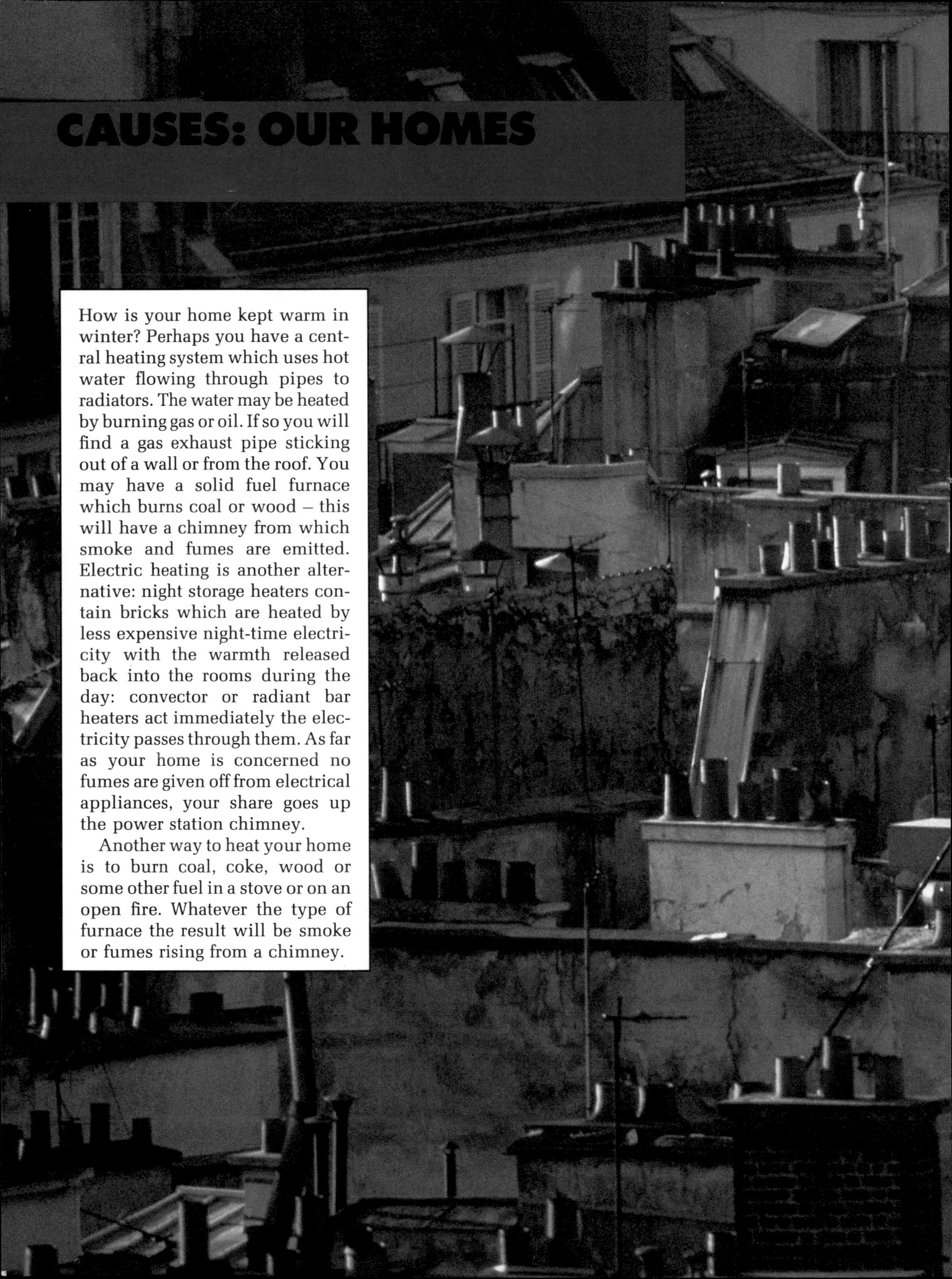

CAUSES: OUR HOMES

How is your home kept warm in winter? Perhaps you have a central heating system which uses hot water flowing through pipes to radiators. The water may be heated by burning gas or oil. If so you will find a gas exhaust pipe sticking out of a wall or from the roof. You may have a solid fuel furnace which burns coal or wood – this will have a chimney from which smoke and fumes are emitted. Electric heating is another alternative: night storage heaters contain bricks which are heated by less expensive night-time electricity with the warmth released back into the rooms during the day: convector or radiant bar heaters act immediately the electricity passes through them. As far as your home is concerned no fumes are given off from electrical appliances, your share goes up the power station chimney.

Another way to heat your home is to burn coal, coke, wood or some other fuel in a stove or on an open fire. Whatever the type of furnace the result will be smoke or fumes rising from a chimney.

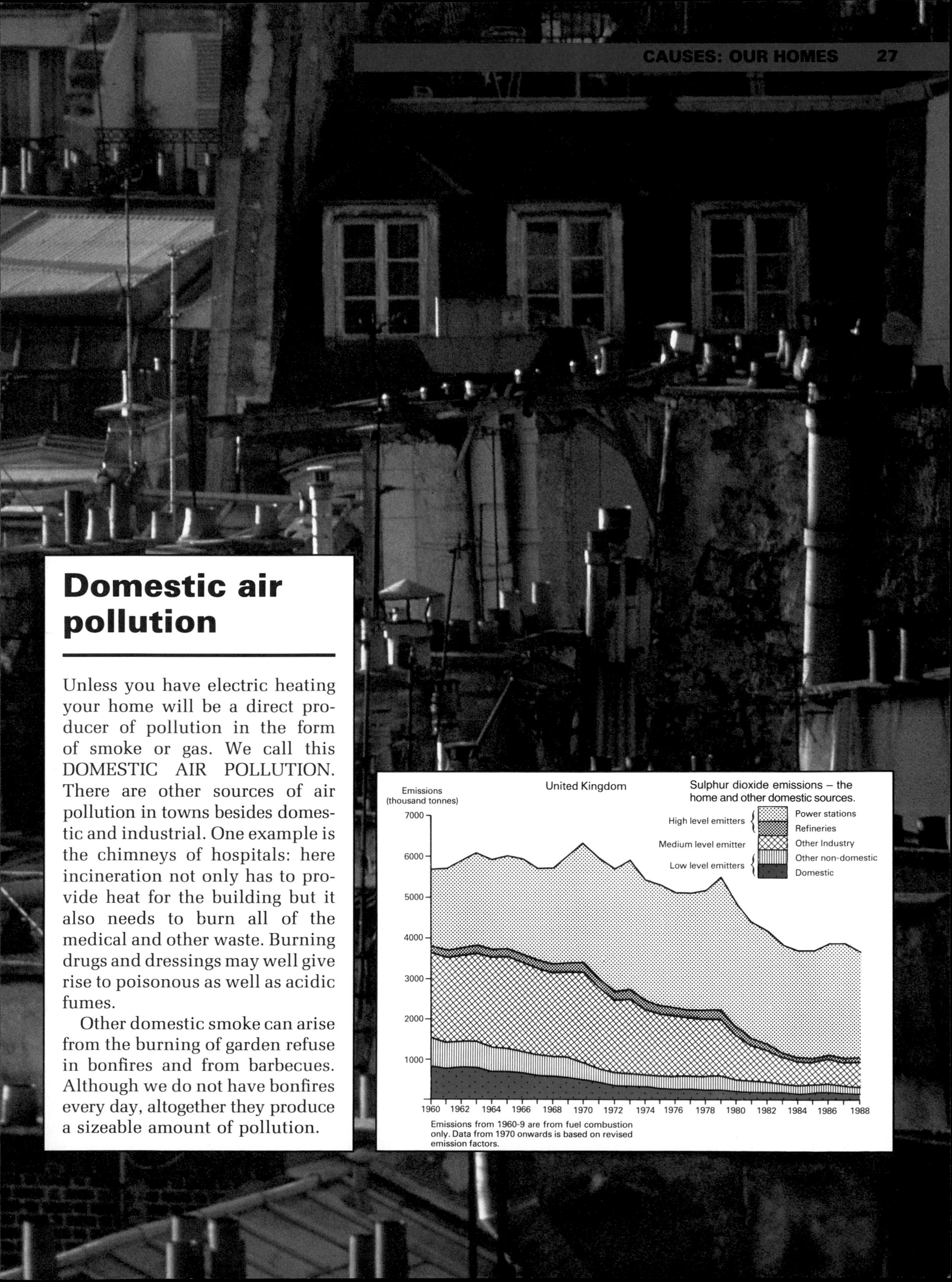

Domestic air pollution

Unless you have electric heating your home will be a direct producer of pollution in the form of smoke or gas. We call this DOMESTIC AIR POLLUTION. There are other sources of air pollution in towns besides domestic and industrial. One example is the chimneys of hospitals: here incineration not only has to provide heat for the building but it also needs to burn all of the medical and other waste. Burning drugs and dressings may well give rise to poisonous as well as acidic fumes.

Other domestic smoke can arise from the burning of garden refuse in bonfires and from barbecues. Although we do not have bonfires every day, altogether they produce a sizeable amount of pollution.

Waste disposal

Domestic waste is collected for disposal: much of it is tipped into landfill sites but a lot is burnt in incinerators to cause more polluted air. The waste itself burns away with little added fuel, to leave only unburnable items such as metal and glass, cinders, ash and smoke. Although we shall look at the way that sulphur emissions from burning fossil fuel can be reduced, it may be convenient here to mention that one of the obvious ways to burn less fuel is to use the heat provided by the incineration of household rubbish to create electricity. There are two methods by which this can be done. The first is simply to burn domestic rubbish in waste disposal plants and use the heat to run a power station built on the same site, rather than let it escape as hot gases up the chimney.

Refuse derived fuel pellets.

But there are difficulties. The whole complex needs to be built at the same time, which requires a great deal of long term planning but it does provide a great opportunity for building a Combined Heat and Power station, something we will discuss later.

Instead of burning the waste, a second way is to convert it into fuel pellets. There are several waste disposal plants doing this already: the illustration shows the sort of product one obtains. These pellets can then be used as a fuel for industry or for a power station either alone or mixed with coal (providing that the pellets have a low sulphur content themselves).

Sewage effluent from urban industrial areas contains not only human waste but the residues from factories. The chemicals in the latter mean that the sludge and solid waste which remains after treatment at water purification works is too toxic (poisonous) to be used as agricultural fertilizer. The only means of getting rid of sewage sludge of this kind is to dump it in appropriate sites or it can be burnt, creating more fumes for the atmosphere to absorb and more acidity to be deposited.

A hospital chimney. In the United Kingdom hospitals are currently exempt from local smoke control laws.

DAMAGE TO BUILDINGS

As you walk around built-up areas look out for damage and decay. Some will be the result of storm damage, but do you notice in old stone buildings – churches for instance – places where the stonework is stained black, where flakes of stone are loose and liable to fall, where worn-away holes appear in the stone and particularly where the faces and other parts of the ornamentation have lost their outline? The photographs on these pages show this sort of damage: it has been brought about by acid rain.

Most stone building blocks are made of limestone or sandstone, both of which are fairly easy to shape or to carve. The lime which makes up the limestone and the lime which binds small particles of sand together to make the sandstone are both affected by acidity.

The reaction to a direct attack by SO_2, or sulphur as dry deposition is common in built-up areas due to local emissions of pollutants. Acid rain, on the other hand, will affect buildings which are well away from sulphur sources. Crystals of salts will form as a result of acid rain attack. Crystals 'grow' and in so doing force the stone to split. Any cracks formed in the rock will fill with rain water

and the ordinary forces of freezing and warming will lead to expansion and contraction, with the end result that lumps of stone fall off. If you look around the bottom of stone walls you will find bits and pieces of broken rock. On a cold night you can actually hear the cracking taking place. This freeze/thaw process is a natural one of rock weathering – but the added effect of Acid Rain speeds it up.

Buildings old and new

Measurements have shown that 30mm of stone has, on average, disappeared from parts of the outside walls of St Paul's Cathedral in London during the last 250 years – of this, in some parts, two-thirds has been lost in the last 40 years. Portland Stone, a particular type of limestone, was used to build the cathedral. Presumably the weather has not changed too much in the last 250 years, so that the quicker pace of destruction must be due to other causes – the strongest possibility is acid attack.

Destruction of stonework

Gypsum (calcium sulphate) is formed when sulphurous products react with lime [see page 52]. Chemically, the sulphur, sulphur dioxide or weak sulphuric acid, react with the calcium carbonate (lime) to form calcium sulphate and carbon dioxide.

The gypsum causes the rock face to flake, which exposes fresh stone for the next attack. Dirt and soot in the air will combine with the gypsum to form the black stain and crusty material, especially where the stone is fairly sheltered.

Some famous afflicted buildings

Athens	The Parthenon
India	The Taj Mahal
Rome	The Colosseum
Washington	Lincoln Memorial
Ottawa	Parliament Building
London	Houses of Parliament
Cologne	Cathedral
Pisa	The Leaning Tower
London	The Tower of London
Venice	St Mark's
Oxford	Sheldonian Theatre

Although famous buildings get the most publicity, ordinary houses and other structures are affected badly. If you are attending an old school for daytime lessons or adult education classes, look around to see what acid damage you can find.

But it does not have to be an old building. Modern buildings made of ferro-concrete can be affected. Ferro-concrete structures are built with concrete blocks reinforced with steel girders and steel mesh. Look and see if you can see rust staining on the walls of such a structure. This means that the steel inside is being attacked chemically. You will observe cracks in the concrete and even lumps breaking away. These both indicate that acidic chemical reaction is taking place. Rust occupies more space than the original ironwork. This expansion causes the concrete to crack and break away. Many modern road bridges and motorway interchanges are decaying in this way. The road salt put down in icy weather also gets into the cracks and causes corrosion (the rotting of the building material). Perhaps the most famous interchange in the United Kingdom is Spaghetti Junction in Birmingham. The scaffolding which often surrounds the supporting columns indicates that all is not well with the structure.

It is true that a building which is very exposed to the weather will show signs of damage which are absent from parts of the same building protected from the worst intensity of the wind and the rain. Sometimes the actual level of pollution will vary so that the lower floors of a building may be exposed to greater pollution than floors higher up.

Stained glass

One aspect of old or religious buildings which gives rise for concern is the way in which acidity is destroying coloured stained glass. The windows of cathedrals, churches and other buildings are being attacked by pollutants which wear away the surface of ancient stained glass.

Take Cologne Cathedral as an example: it contains about 10,000 square metres of glass. The glass has been reduced in thickness from its original 6mm to 2mm over most of its areas, while the fired-on black paint has virtually gone altogether. Inside, the coloured paints are well on their way to being destroyed by the condensation caused by heat changes – outside, acidity has removed almost all of the paint. Cracking of the glass has taken place on nearly all of the individual pieces: this is not the fault of acidity but of wind pressure, the removal of the glass in two world wars and the sonic bangs of jet aircraft.

To combat all of this restorers remove the old glass, clean it, glue the cracks, cut away the old lead which held the individual pieces together and reapply the black paint. The restored piece is then embedded in a yielding mass of acrylic film and sandwiched between the two thin sheets of transparent glass. New strips of lead hold this together to create a piece of glass back to its original 6mm thickness. The method is expensive but after treatment the glass glows again with the splendid colours of the original.

The Statue of Liberty in New York Harbour has been restored recently. Its copper cladding was eroding away in some places, caused by a mixture of sea salts and Acid Rain.

EFFECTS ON LAKES AND RIVERS

The day is perfect, pleasantly warm but not hot; the wind is still, the sun shining, apart from the few occasions when cotton-wool clouds – the cumulus variety – temporarily come between the sun and the earth. Fir trees stretch along the banks of a large lake and, together with those few clouds, are reflected in the calm crystal clear waters below. This is what has drawn the visitor to Sweden, to the fresh clean air, the beautiful scenery and the magic of the lakeland scene – perfect, absolutely perfect. But is all as perfect as it may seem? Unfortunately the answer is NO! Invisible in the water vapour of the air is the acidic pollution – that wonderful crystal water is clear only because it contains no life. The millions of minute plants floating free in a 'happy' lake are not to be found, neither is the animal life both great and small; no fish, no larvae, nothing. The cause – Acid Rain formed from sulphuric

pollutants from the chimneys of Britain, Germany and the other countries surrounding the Scandinavian Peninsula: and from Swedish industry too. The reason – the water has a high acidity (pH 4.5) at least 1000 times as acidic as it was in the years before 1940.

The perfect scene is just an illusion of splendour and although walkers, cyclists and car travellers tour the area, the anglers are gone and the hotels which catered for them are shut.

This did not come about suddenly as it did with the notorious chemical spill into the river Rhine. There were no floating bodies of fish or eels, no smell of rotting corpses, just a gradual decline into a lifeless expanse of sparkling water – and no idea of any link with the discharge of sulphurous emissions from power stations hundreds of miles away in another country!

Llynnbrianne, Wales

Acid Rain: what proof?

The real problem is to prove that the acidity of the falling rain and snow does increase the acidity of the lakes and rivers. With the constant change of the water as it flows toward the sea the extra direct rain falling on it is of little account. Most water which feeds the rivers and lakes comes from the rain and melting snows in the mountains, and from the rain which has percolated through the soil. The rain washes down through the trees and plants and falls on the ground below: it gathers chemical matter from the vegetation and the land surface: as it soaks through the earth and rock it absorbs minerals through several different sorts of chemical reaction. This is a series of complex processes very difficult to untangle and even more difficult when it comes to identifying cause and effect.

The amount of pollution will depend on:

1 The amount of rainfall.
2 The size of the catchment area.
3 The sort of rocks beneath the soil. Acidic rocks will increase the acidity of the water draining through it.
4 The use made of the land. Ploughed land will enable chemicals to be easily leached (extracted) from the soil. Vegetation cover (trees, heather, grass) will protect the soil. Urban areas will add pollutants to draining water.

How can anyone say for certain 'this drop of Acid Rain caused this drop of acid water'?

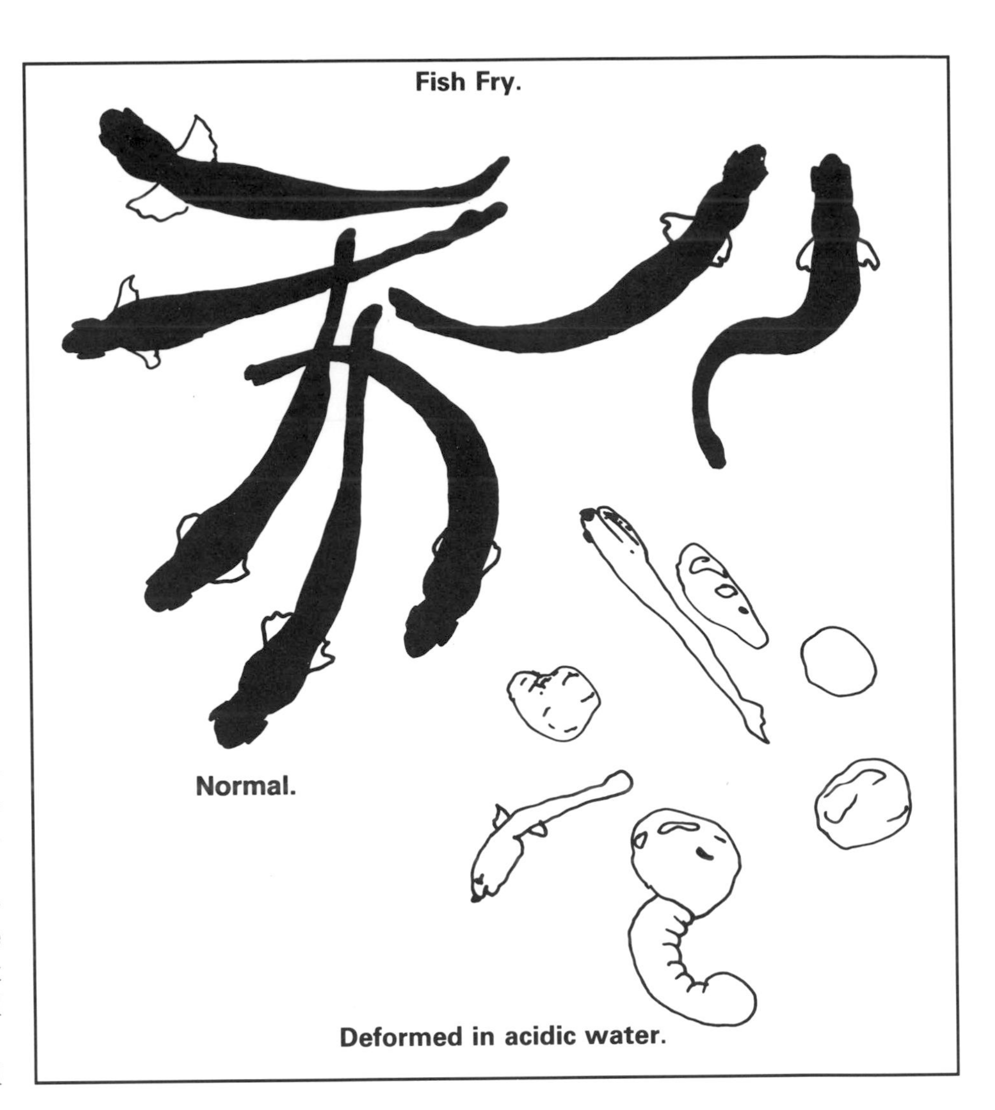

Effect on fish

Experiments show that acidic water causes fish fry to be deformed. The fry, the young of fish, develop from eggs spawned by female fish. In acidic waters fewer eggs hatch and those that do often fail to grow. Salmonid species (trout and salmon) are easily affected by acidification and as they are the main quarry of anglers, the adverse results are soon evident. The 'tell-tale' sign is that anglers catch one large fish after another as there are fewer and fewer small fish swimming in the water: only the older, larger and slightly more resistant fish are left.

Changes in the body salts of fish are another effect of high acidity. The calcium in the bones is attacked and the skeleton of the creature is unable to withstand the pull of its muscles. As a result fish become deformed and eventually die.

Acid Rain percolating through the soil releases metals into the water. One of these, aluminium, has a terrible effect on fish: it causes mucus (a gummy or slimy substance) to clog the gills, makes breathing difficult and so causing death. Sometimes fish 'snort' in an attempt to clear the mucus – these 'sneezing' fish are a sure sign of serious acidity.

Acid surge

Fish and other large creatures can adapt to lower levels of pH, provided it does not go too far. What they cannot tolerate is a sudden increase in acidity. Snow melt and storm episodes may result in a more than tenfold increase in acidity, disastrous to the fish.

Most aquatic animals reproduce in the shallower waters around the edge of a lake or alongside the shallower sections of river bank. The diagram on page 13 indicates that a lake may not have the same pH valuc across its whole width. The bankside area known as the *Zone of Reproduction* (pH values down to 4.5) is most affected by acid surges, as the snow melts first into the bankside water. Incoming rivers discharge into that area also.

Extra nutrients are brought down by the spring thaw which attracts the fish where they become prey to the extra 'poison' of reduced pH. Although it is the most dangerous place for lake life it is the place where most creatures collect: unfortunately fish do not know the risk they take.

Remember, many larger creatures depend on food from rivers and lakes – no food leads to no kingfishers, herons, otters or beavers. A dead lake means a surrounding area deprived of much of its natural life.

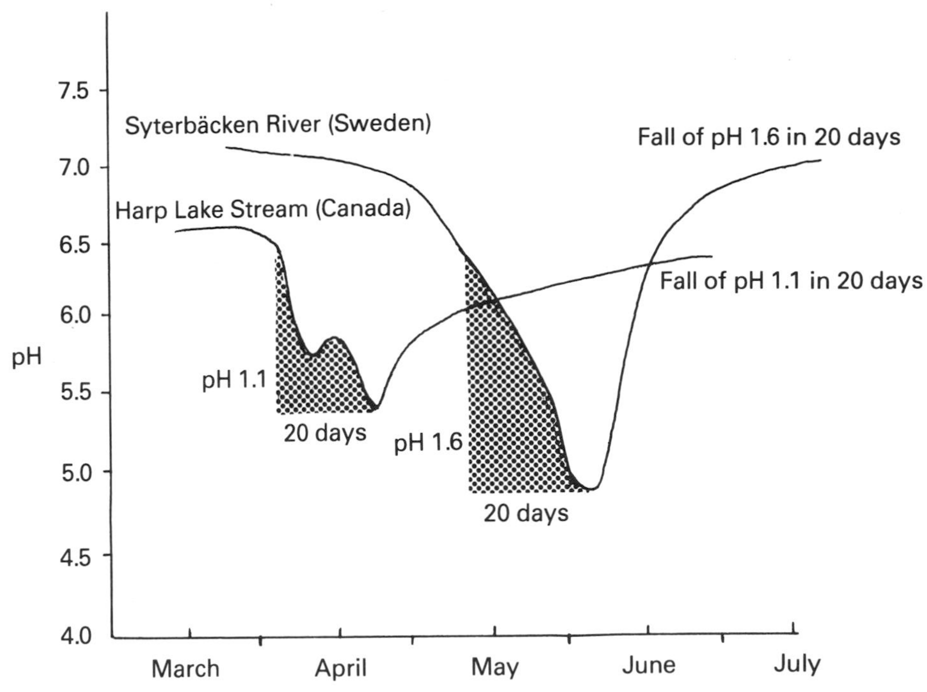

Graph to show the drop in pH value of the water at the time of acidic surge.

Plain common sense

Yet when all of the uncertainties have been expressed; when all of the experiments and observations have been carried out and no firm conclusions have been reached; when all of the 'not proven' statements have been spoken, who can really doubt that high acidity in the rain is linked to the pollutant emissions from chimneys and from vehicles, and that dead and dying rivers and lakes are the result. As we often say when matters are not that certain 'It's just plain commonsense!' Unless we do something to stop the emissions, more and more lakes will look 'perfect'.

EFFECTS ON TREES AND PLANTS

Trees such as oak and ash have broad leaves which fall every autumn, causing the branches to be bare throughout the winter until the new spring encourages a fresh set of leaves to grow – these are the DECIDUOUS trees. Other trees have small pointed leaves, known as needles. These fall from the tree a few at a time throughout the year. As the trees are always covered with needles they are ever-green – these are the CONIFEROUS trees.

Trees are prone to attack by insects, drought, fungi and disease: often more than one sympton of ill health is present. It is difficult, therefore, to isolate any one cause as being the main reason for the decline of a tree: it is particularly difficult to assess the influence of Acid Rain on trees. However, the usual tree attackers give rise to recognisable symptoms ... since the middle of the 1970s trees have begun to show other symptoms

(a) In conifers: the thinning of the top of the centre of the tree known as 'crown die back'; needles turning yellow and falling off; bark damage; the stunting of growth and root damage.

(b) In deciduous trees: leaves which are discoloured; leaves which have the wrong shape; crown die back; early fall of leaves; bark damage; loss of small roots and, particularly, fewer leaves which means that more daylight can be seen through the branches.

Both types of tree appear to have become less resistant to other attackers; in time the death of the tree results: where the trees are part of a forest large areas, or stands, of trees die. Despite the fact that it cannot be stated conclusively, all of the evidence suggests that acidity in the air plays a decisive part in this extra damage. The Germans have given this loss of forest a special name: the *Waldersterben*.

Experimenting on conifers

Other causes

There is no doubt that local conditions affect healthy tree growth. These are some of them:

- the height of the ground above sea level
- the amount of wind which blows into the tree
- the sort of soil around its roots
- the depth of the soil
- how close the tree is to pollution (from industry, vehicles or domestic sources)

All of these factors affect how well trees grow. Only part of any damage is the result of acid deposition.

How are trees affected?

There are several theories on the way that Acid Rain affects trees. It may be that the acid leaches certain minerals from the soil which are essential to the healthy growth of the tree. In particular this applies to magnesium where deficiency may cause the yellowing of leaves and needles. Laboratory work shows that magnesium is essential to the structure of chlorophyll, which is the chemical responsible for the green colouration of leaves.

Secondly ozone, which results from the increase in NOx emissions associated with vehicle exhaust, damages the surface of pine needles and has been found to damage broad-leaved trees such as the oak, ash, beech and birch. It has been suggested that ozone causes oak trees to dry out and die. The most likely explanation is that ozone adds another stress to trees affected by drought or other damage.

Thirdly Acid Rain adds to the natural acidity of the soil, especially where trees are growing on soils derived from granite and have root systems in more acidic conditions than those growing on limestone. The extra acidity tends to cause the aluminium in the soil to become concentrated and attacks the fine root system, which prevents the absorption of water and nutrients – thus harming the well being of the tree. Another theory suggests that extra nitrogen itself may be causing tree growth to be too rapid, which leads to weakness and increased susceptibility to attack from other injurious factors.

Percentage of trees showing more than 10 per cent defoliation (loss of leaves).
Except where underlined the figures are from nationwide surveys. * indicates percentage of conifer trees only. (Source: United Nations Economic Commission for Europe/International Cooperative programme, 1989.)

How are plants and crops affected?

It can be argued that acid deposition is helpful to farmers since they have to put nitrogen onto the soil and use sulphates to improve crop yield: it all depends on the fertility of the soil in the first place. Sulphur compounds are also used to control fungal diseases such as black spot on roses.

Nevertheless it is generally recognized that acid deposition does reduce the yield of crops. The US Environmental Protection Agency has found that Acid Rain can damage the leaves of spinach, lettuce, beans and radishes. The yield from carrots, cauliflowers, beet and mustard was reduced. These findings resulted from experiments where the high levels of acidity were unlikely to be found in field conditions. Although there is a general feeling that Acid Rain does reduce yields, there is no conclusive scientific evidence to support this belief. Certainly there is evidence that crop yields are increasingly reduced the nearer the growing area is to an industrial zone. The US National Crop Loss Assessment Network has investigated crop losses from SO_2, NOx and ozone in wheat, corn, soya beans and peanuts. Losses are estimated at over $3000 million annually in the early 1980s and about $2000 million annually in the mid-1980s.

Acid Rain releases metals in the soils and makes them available for take up by growing plants – the cabbage family is particularly vulnerable. In 1990 in Czechoslovakia tests by the State Veterinary Council showed that 20% of potatoes, 5% of other vegetables and 5% of beef had more than the permitted level of cadmium: lettuces had 15% over the permitted nitrate level.

Ozone at ground level, caused by the photochemical action of sunlight on nitrogen gases, is injurious to plant growth. It attacks the cells of leaves, destroys the green matter (chlorophyll), thus lowering the rate of photosynthesis. This leads to dark-coloured patches appearing on the leaves, often put down to weather conditions but now proved to be ozone damage.

ACID RAIN AND OUR HEALTH

Fresh air, pure water and clean food are the essentials for a healthy body and a long life. If any of these are adversely affected by pollution, the human body will suffer to a greater or less extent.

Smog in Los Angeles showing the effect of the inversion layer.

Air

It is difficult to isolate one pollutant from another as we breathe air and not individual gases. Medical science knows that SO_2 and the resulting acidic moisture can harm lung tissue and cause many breathing problems. Most people are not

in the unfortunate position of having to breathe poisonous air on a daily basis, but many of us are exposed to low levels of acid pollution.

In 1989 the overthrow of the repressive communist governments in Eastern Europe not only led to enormous changes in political power but also revealed the true state of pollution brought about by badly run power stations and industrial processes. In Romania doctors at a clinic in Copsa Mica, in the industrial area of Transylvania, were reported (*Environment Guardian* 19 January 1990) as saying 'Levels of cancer, skin disease, respiratory problems, hypertension (high blood pressure), and premature births are all higher than average here. A man will come in with bronchitis but we have real trouble diagnosing it because his problems are compounded by so many other diseases.' The dry deposition of pollutants from the petroleum processing plant, the lead smelter and the carbon black factory have turned Copsa Mica into a 'chemical plague town', where the snow of winter is more commonly black than white. These factories emit a mixture of carbon powder, sulphuric, oxalic and formic acids, as well as lead and zinc. In particular the pollution levels are very high within a 30 km (19 mile) radius of the industrial centre. In eastern Germany at Cottbus near the Polish and Czechoslovakian borders the burning of lignite (brown coal) by all three countries has contributed to the death rate among men being twice the European average. Bronchitis, asthma and other lung disorders account for most although the increase in cancerous tumours is alarming.

It is not necessary to go to Eastern Europe to find instances of air pollution having a devastating effect on human health. The photochemical smogs of Los Angeles, Athens and other urban areas are the cause of lung disorders. In every country there is some organization aimed at improving the situation – in the USA the American Lung Association and in the UK the Clean Air Society are two examples. The term smog was originally given to the mixture of fog and SO_2 emitted from house chimneys. It was the London smog of 1952 which caused about 4000 people to die from lung and heart illness which finally brought things to a head in Britain and led to the Clean Air Act of 1956.

Water

Water is also essential to our well-being – since almost three quarters of our body is made up of water this is understandable. We take in water directly as a liquid or as the main constituent of other liquids such as orange juice, coke, or a cup of coffee. We also take it in with the food we eat: for example lettuce is almost entirely made up of water – let a lettuce leaf dry out and you will see how little solid matter is left. If water is

Californian schools are issued with instruction books to tell them what to do when smog occurs.

highly acidic it will carry heavy metals into our bodies. We have already seen that Acid Rain causes the release of aluminium and cadmium from the soil which may be taken up by plants which we then eat. Aluminium is thought to be associated with Alzheimer's disease, which leads to loss of memory and eccentric behaviour in elderly people and cadmium is poisonous in very small quantities. Acid water can also cause the release of metal used in pipes and storage tanks in the home. Copper pipes are common in the modern home and despite the fact that plastic water storage tanks are now the normal installation many millions of metal tanks are in use, commonly made of galvanized iron (iron covered in zinc) or less commonly stainless steel. Hot water tanks are usually made of copper.

Water flows through the pipes and rubs against the sides, causing minute particles of the metal to be dissolved. Tests have shown that the lower the pH the greater the absorption. Where the water is hard, because it contains more lime, then the pH is not so low: where the water is soft the acidic effect is greater. Always allow water to run for a little while first thing in the morning for metal absorption is greater where the water has been standing for a long time: it is also true that warm water dissolves more metal than cold. Swedish doctors are convinced that there is a direct link between acid water, copper pipes and diarrhoea illness with babies, while lead from old-fashioned water pipes is known to cause brain damage in young children.

PREVENTION: TREATING AND CHOOSING FUEL

To reduce Acid Rain industry's main concern is to reduce the amount of SO_2 being produced when fossil fuel is burnt. Therefore they can:

use coal which contains little sulphur
or remove the sulphur which is in the coal
or use another type of fuel
or burn the coal in such a way that the sulphur is destroyed

Burn low sulphur coal

Coal varies in the amount of sulphur it contains; this can be low, medium or high, anything between a half per cent to 5 per cent. At the worst this means that up to one twentieth of the coal which arrives at a power station is sulphur and not coal. Hard coal such as anthracite generally contains less sulphur than softer coals – for example bituminous coal and lignite (brown coal). However German brown coal from the Rhineland has only one per cent sulphur whereas United Kingdom coal has an average of about three per cent. One coal seam may have more sulphur in it than another seam at the same mine.

Unfortunately the idea of burning low sulphur coal is not as simple as it sounds. Transport costs can mean that imported low sulphur coal costs more than local coal. In the USA it costs so much to move low sulphur coal from the west to the industrial east that local high sulphur coal is used. To move coal around means more pollution from transport vehicles and more disturbance for people on the route. Another snag is that furnaces using high sulphur coal need to be expensively altered to deal with the low sulphur variety. But perhaps the greatest difficulty is that local American, British or other miners become unemployed and their mines closed down.

Clean the coal by removing the sulphur

One way that sulphur is contained in coal is as Iron Pyrites. As iron it is heavy and attracted to magnets. Coal can be crushed into small pieces, mixed with water and passed through a tank so that the pyrites sink to the bottom leaving the coal to be removed. Or else the powdered coal can be passed under powerful magnets which attract the pyrites away from the fuel. Either method reduces the SO_2 by about a third. Electrostatic precipitators (see page 20) can be used to attract the coal particles while the heavy pyrites and other waste falls below. Experiments in the USA with an electrostatically charged revolving cylinder have removed up to two-thirds of sulphur and about half of the ash.

However, much of the sulphur in the coal is chemically trapped into the fuel itself. It is possible to remove it but it is more expensive and intricate than removing the pyrites. Chemicals can remove the sulphur (and the pyrites); microwaves or electron beams will rid the coal of organic sulphur, but all of these methods are not yet cheap enough to be commercially possible, despite the fact that up to 90 per cent of the sulphur has been removed in experiments. New processes will hopefully make it economically possible to clean coal of all sulphur before too long.

Cleaning coal

Quite a lot of cleaning does go on, particularly in the USA and Germany. At least 40 per cent of the coal and lignite mined in the eastern coalfield of the USA is cleaned resulting in an estimated reduction of about 2½ million tonnes of SO_2 every year. Simple arithmetic shows that over 5 million tonnes of SO_2 would be removed if all of that coal were cleaned – but what is done in the USA is much better than in most other parts of the world.

Nitrogen oxides (NOx)

It is important to try to reduce the emissions of NOx at the same time as removing the SO_2. Nitrogen gases are formed by chemical reaction when burning fuel in air: by lowering the temperature and the time in which the air is in contact with the incineration there will be a considerable reduction in the formation of NOx. It helps if the air is fed into the burners in different stages rather than all at once. NOx emissions can be cut by half with relatively simple and inexpensive techniques. Equipment costs are low and, although compared with a traditional boiler the new type of furnaces cost more, it is a very small proportion of the cost of the whole incineration system.

Use a fuel other than coal

The simplest cure of all is to use oil or natural gas instead of coal. Crude oil contains up to 3 per cent of sulphur. When refined the light oils can be easily reduced to about a half per cent sulphur content, but the heavy oils can end up at 5 per cent. There are techniques already in wide use to reduce this to a half per cent, but it adds about $50 a tonne to the cost. 90 per cent of the sulphur can be removed by *Direct Desulphurization*, unfortunately not yet in use commercially. Natural gas is very low in sulphur: it has other environmental advantages as, once the gas field has been tapped and the pipeline laid, there are no transport costs or vehicle pollution.

The real cost?

So there *are* ways to remove sulphur before coal is burnt. But they have to be paid for and this results in costlier electricity. The cost is usually measured in money – the real cost is in the damage done by pollution if the SO_2 gets into the atmosphere. No cash price is too high if it solves that problem.

Destroy the sulphur when the coal is burnt

Methods of coal incineration which destroy SO_2.

	Fluidized Bed Combustion (FBC)	Lime Injected in Multistage Burner (LIMB)	Integrated Coal Gasification Combined Cycle (IGCC)
Method	Make coal dust act as a liquid by jetting air into it. Add ash and lime. Burn at 900°C which is half normal temperature.	Slow burn mixture of coal and lime.	Change coal to gas to make electricity via gas turbine. Use waste heat to make steam to drive steam turbine.
SO_2 reduction	90%	50–75%	99%
NOx reduction	50%	50%	95%
Waste	gypsum and ash	gypsum and ash	ash
Special notes	Operating FBC under pressure is more efficient – called Pressurized Fluidized Bed Combustion (PFBC).	Other systems use calcium chemicals especially with brown coal.	The amount of sulphur in coal does not matter. Using two sorts of electricity production gives process its name.

PREVENTION: USE OF LIME

What is limestone?

Limestone is a sedimentary rock mostly made up of the remains of creatures which lived in water, usually the sea. All limestones are chemically Calcium Carbonate. Lime is purified limestone crushed into a powder.

In the battle against Acid Rain lime is used in four main ways. First of all it can be added to lakes and rivers, secondly it can be spread on the soil surrounding acid lakes, thirdly it can be jetted on to the gases which pass up factory and power station chimneys, or fourthly mixed with coal dust in FBC furnaces (page 49).

Lime in lakes and rivers

Powdered lime added to the water of lakes has an almost immediate effect in reducing the acidity. Unfortunately it does not cure the situation – it merely temporarily counteracts it. How long it will be effective depends on the speed at which water flows through the lake: in some lakes one liming will last 15 years, in others for a year only. On average Swedish lakes need extra lime every 4 or 5 years.

Where rivers are concerned the flow of the water makes it more difficult. Systems have been devised which allow small amounts to be added to the water at regular intervals, but not only does this have to be a continuous operation it is difficult to have the machinery react to times when greater flow occurs, for example, when snow melts or storms have drenched the area.

Lakes which have been limed regain their life provided all has not disappeared. The USA, Canada, Wales, Scotland, Norway and Sweden are but some of the countries where lakes have been limed successfully. However the real cure has to be the prevention of acidic emissions – lime is only a temporary corrective measure.

Lime on the soil

Many farmers and gardeners give their soil a top dressing of lime every year to make it less acidic. Liming the soil near a lake has also been found to help counteract the acidity in the lake water and to be effective in preventing the input of toxic metals. This is a long term 'cure'. The reduction in acidity of the water takes a long time but it lasts for longer. However the effect on conifer trees is not so good, as it has been proved that lime reduces growth. Open land can be treated with lime quite easily, but if it is forest covered it may be necessary to lime from the air.

Given time liming does help to raise the pH value of the water passing through the soil and 'fix' aluminium and other metals. Even if new sulphurous Acid Rain does not fall in an area there is still the problem of the sulphur that has built up in the soil over many years.

Gypsum

Calcium Sulphate can be turned into Gypsum by using air oxidation. Gypsum occurs naturally in nature but it is not in sufficient supply (particularly in Japan). Other sources are needed for commercial use. It is best known as PLASTER OF PARIS, which is used to make casts to keep broken limbs in position. It is also used to make plasterboard which is common in new buildings for wall and ceiling cover.

Lime used on chimney gases

This is known as Flue Gas Desulphurization (FGD) with the *lime jetting* method the most common, but not the only method used. FGD is the most widely publicized method for dealing with the SO_2 emitted from the chimneys of power stations and factories – particularly in Japan and the USA. The common process is to spray chemicals on the gases as they pass up the chimney, in an operation known as scrubbing. The chemical, usually lime, absorbs the sulphur to become calcium sulphite or calcium sulphate. Gypsum is made by oxidization of the residue.

1 *Wet FGD* The lime is sprayed into the chimney as a SLURRY – a mixture of lime and water. A chemical sludge, like thick double cream, falls to the bottom, while the 'scrubbed'

Material flows for a 2000 megawatt coal-fired power station with a limestone-gypsum flue gas desulphurization system.

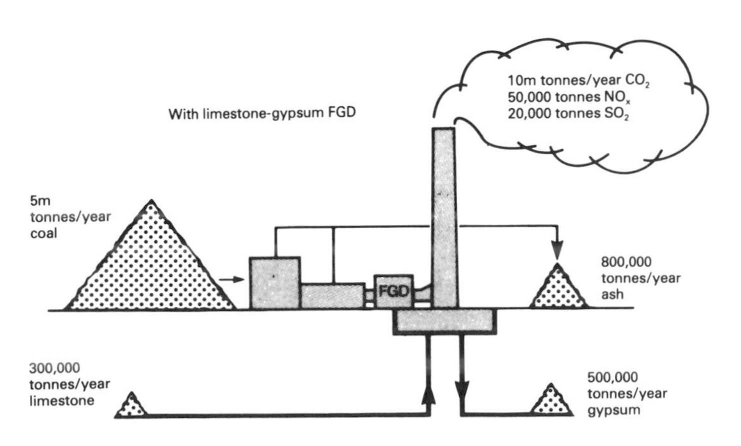

Material flows for a 2000 megawatt coal-fired power station with a Wellman-Lord desulphurization system.

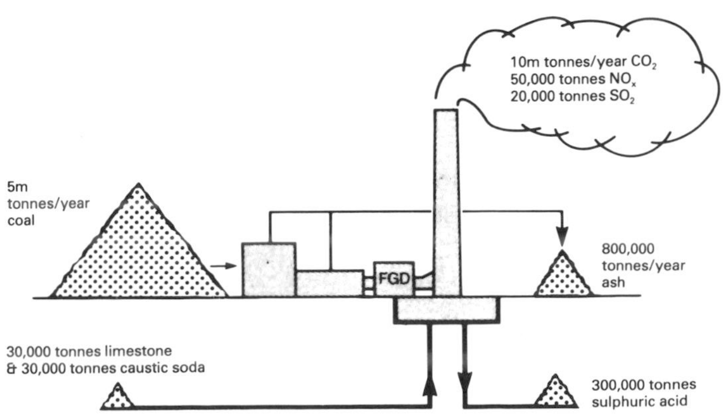

exhaust goes up the chimney into the air. Over 90 per cent of SO_2 is removed but wet sludge has to be collected, dried and treated: this needs electric power to add to the extra cost involved in FGD.

2 *Dry FGD* The chimney gases are sprayed with either ammonia to form dry pellets of ammonium sulphate (a fertilizer) or a lime slurry which is dried at the same time as it absorbs the SO_2. The dry methods remove less of the SO_2 but are cheaper to run, need fewer workers, have less production problems and produce no wet sludge.

3 *Regenerative Method of FGD* The chemicals which remove the SO_2 are recirculated. One example is the Wellman-Lord system. In this sodium sulphite is circulated in the flue gases to form sulphur or sulphuric acid, depending on the design. With the sulphur it is necessary to use natural gas as well. With a 2000MW power station producing 300,000 tonnes of acid every year, disposal becomes a problem – but at least it can be controlled rather than fall as Acid Rain anywhere on the land below.

If anti-pollution methods are added after a power station has been built the process is called RETROFIT.

Other FGD processes which do not use lime are at the experimental stage.

1 *Electron Beam (E Beam)* This imitates the formation of Acid Rain in the chimney. Flue gases are sprayed with ammonia and water: the mixture passes through an electron beam; acids are created; ammonia reacts with the acids to form chemical compounds suitable for use as fertilizers. An E Beam system is installed in a Tennessee Valley Authority Power Station in the USA.

2 *Copper Oxide (CuO) FGD* CuO is used instead of lime to react with the SO_2 to form copper sulphate. If ammonia is also used the NOx is reduced. CuO FGD is still being developed.

3 *SO_2 filter using an electrochemical cell* The negative electrode attracts the SO_2 from the flue gases before they reach the chimney. Oleum, a highly concentrated type of sulphuric acid, is formed. Laboratory tests remove 99 per cent of the SO_2 at 20 per cent of the cost of other methods. NOx is also removed. This filter system is still experimental.

The disadvantage of FGD

1 Disposal of the by-products – industry can only use so much gypsum or sulphuric acid.

2 Limestone has to be quarried from some of the most scenic countryside which spoils the natural beauty; creates noise and dust; increases lorry traffic – one environmental problem is made in order to remedy another!

PREVENTION: VEHICLE EXHAUST

We can cut down on vehicle pollution if we use motor vehicles less, drive more slowly and carefully so as to use less fuel or use public transport. If we do none of these things we can still cut down on pollution by better technology. There are four main ways:

1 Use CATALYTIC CONVERTERS
2 Have LEAN BURN ENGINES
3 Use NON-POLLUTING FUEL
4 Use NON-METAL ENGINES

Catalytic converters (cats)

A catalytic converter looks like an exhaust silencer. If you cannot recognize one of these, observe a motor cycle exhaust pipe: just before the end there will be a bulge through which all exhaust gases have to pass so that the noise is reduced. A cat is similar except that it deals with the gases.

If the converter is working well, only about 10 per cent of the original pollutant gases will pass into the atmosphere. BUT it does have to be working well. In order to make sure of this a check system has to be organized in countries where the use of cats is law.

Catalytic converters are made completely inoperable by lead. This means that unleaded petrol has to be available. One of the difficulties when catalytic converters were made a legal requirement in the USA was that leaded gasoline (petrol) was cheaper than unleaded. The temptation to use the cheaper variety (even though it ruined the converter) was overwhelming with many car owners. In the UK the government charges less tax on unleaded fuel in order to make it cheaper and so encourage motorists to use it.

Catalytic converters are relatively expensive, adding about £300 ($500) to the cost of a car. Their life span is about 50,000 miles which for many commercial users would mean a new converter every year.

Cars depreciate in value very quickly – in other words their resale value goes down. To pay an extra £300 on a car which costs many thousands of pounds when new is quite a different matter from paying another £300 for a new converter when a second-hand car has only cost a few hundred pounds.

At first it was necessary to have several converters, each to deal with a pollutant. Now all of the gases and hydrocarbons are changed

A car with a three-way catalytic converter.

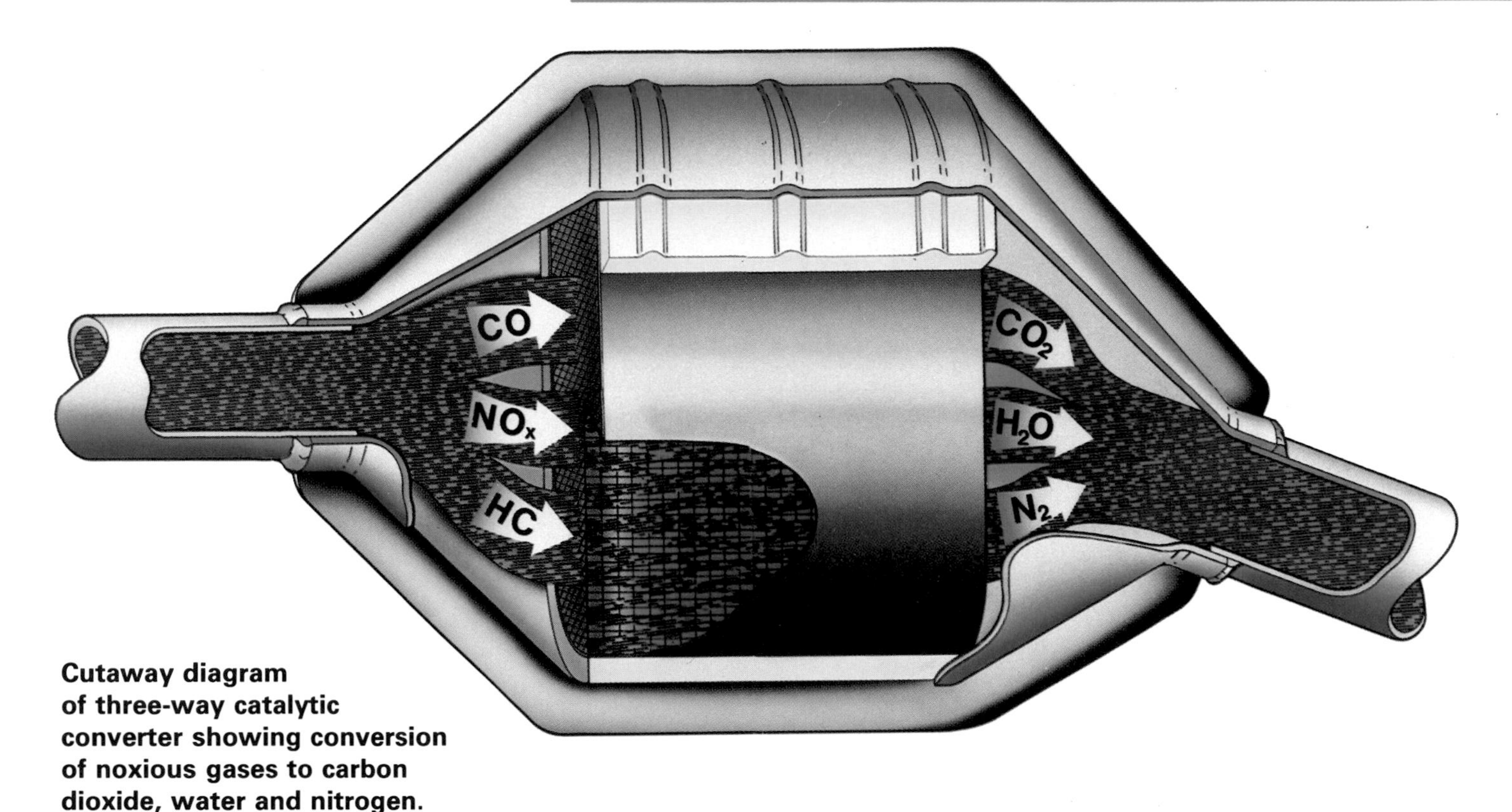

Cutaway diagram of three-way catalytic converter showing conversion of noxious gases to carbon dioxide, water and nitrogen.

in one THREE WAY CATALYTIC CONVERTER. Expensive metals are used – platinum, for example, is a metal that is also used to make jewellery. It is only possible to use these metals from an economic viewpoint because they act as CATALYSTS. A catalyst is a chemical which causes a reaction to take place between other chemicals but which is itself not altered. This means it is possible to recycle the precious metals from outworn converters.

Cats are mandatory (enforceable by law) in countries such as Japan, the USA and West Germany – but not yet in the United Kingdom for all vehicles.

SMOG CHECK

Vehicles with pollution control devices need regular checking.

Lean burn engines

The lean burn engine is one where the fuel is burned in the presence of relatively large quantities of air, about 18 to 22 times the amount of fuel. It is the burning of the fuel which causes the explosion inside the combustion chamber of the engine. This forces the piston down the cylinder and causes the drive shaft of the vehicle to revolve. After the explosion some of the fuel remains together with the gases which are formed when the burn occurs. All of these are expelled from the engine via the exhaust system and on into the air. The graph shows that at one point the air-to-fuel ratio is such that when the explosion occurs all of the fuel is burnt. This is called the STOICHIOMETRIC RATIO; it varies slightly with air temperature and the quality of the petrol used. Unfortunately it is at this point the formation of NOx is the highest although CO_2 and HC are at a fairly low stage. Reference to the graph will show that the best compromise is reached when the mixture is weak in the 18:1 to 22:1 range – the lean burn mixture level.

It is obvious that engine design is critical when using special techniques. In the case of the lean burn engine it is claimed that minimal changes in engine complexity are necessary. The basic idea is to increase the amount of turbulence, that is the swirling around of the petrol and air just before and during the explosion, together with increasing the area where the electric spark is applied to make the mixture ignite. Both of these 'extra' features cause very fast burning of the fuel/air mixture. This is of particular importance in preventing too many hydrocarbons being produced.

Stoichiometric ratio
Nitrogen Oxides
Hydrocarbons
Carbon Monoxide
10 12 14 16 18 20 22
Rich-Burn Lean-Burn

General effect of air–fuel ratio on exhaust emissions.

Non-polluting fuels

Electric battery vehicles are not polluting. The difficulty with the electric vehicles is that the batteries need to be recharged about every hundred miles – not much use for distance travel. Hopefully the development of the fuel cell will change this for the better. Solar powered engines are possible. Canadian researchers are confidently forecasting they will have an electric vehicle powered by aluminium and air running economically in this decade. Using a new system which combines conventional batteries and new aluminium-air fuel cells, researchers hope to extend distance to at least 200 miles without recharging.

Some fuels are less polluting than diesel oil or petrol. Ethanol is an alcohol made from vegetation and is in use in some countries, but it is not as efficient as petrol. Hydrogen is perfectly clean as a fuel, for it produces water as a waste product. It is used for rocket motors – but it costs a lot to produce and is very explosive and dangerous to control. However, scientists are working on hydrogen as a fuel, and it is quite possible that it will be available for general use early in the next century.

Non-metal engines

To go further on the same amount of petrol would have a similar effect as using better fuel. This may be possible shortly with cars which have ceramic (cups and saucers are ceramic) engines and not ones made of metal. These may permit 200 miles to every gallon of fuel. Ford have already developed an engine where most of the component parts are made of plastic, which makes it much lighter and more efficient.

Cats v lean burn

Advantages	Disadvantages
Cats	
Can be used on old or new vehicles Fits any vehicle Eliminates nearly all 'acid' gases	An extra piece of equipment Needs replacing Creates CO_2 – a Greenhouse Gas Less efficient when old Engines need adjustment Must use unleaded fuel Needs regular checking
Lean Burn Engines	
Relatively simple Not an extra cost Can use any petrol type 20 per cent improvement in fuel consumption claimed Operates for whole life of vehicle No extra expense for replacement Not an extra piece of equipment Vehicle speed has little effect on emissions	Only certain models have lean burn – customer choice restricted Older vehicles cannot be adapted Still makes some pollution Engine needs careful adjustment

WHAT CAN I DO?

I despair as I pass a coal-fired power station and see the smoke pouring from the chimney. I know what its pollutants are doing to the atmosphere and the effect they are having on trees, plants, buildings, lakes and people far away from the smoking stack. There seems to be nothing I can do to improve the situation: but then I remember the rows and rows of disused chimney pots in London and in other urban areas. Forty years ago they too were belching smoke into the air. Then ordinary people decided they had had enough of the smoke-laden fogs, blackened buildings and hospital wards full of bronchial sufferers. Politicians got the message to do something about it – and they passed the Clean Air Act.

So why not persuade the people who make the laws in every country to tackle the problem of noxious emissions? One way is through the ballot box, where it is possible to vote into power the politicians who believe in reducing pollution. But not everyone is in a position to vote or may be somewhere which lacks candidates who 'think green'. In this case political opinion has to be changed. In a democracy this is done by peaceful *persuasion*. No one is too young or too unimportant to make their opinion heard.

Energy conservation

Persuasion may seem too long-term and rather remote from our day-to-day life – so is there something we can do now? At the beginning of this book I said that each of us 'owned' a part of the smoke which rose from the power station or from industry, or indeed the exhaust fumes spurting from the truck or lorry. Why? Because the power station is making electricity for us to use, the factory is making something which is eventually part of our personal property and the road traffic is moving something for us to eat or use. Perhaps this is too simple an explanation but if we cut down on many of the things we use we can also cut down a little on pollution. We call this ENERGY CONSERVATION.

Energy conservation is certainly the name of the game where Acid Deposition is concerned. Already we know that much of the extra acidity is due to the SO_2 from fossil fuel power stations, especially those that are coal-fired. It is true that some politicians who wish to find an excuse for not taking unpopular measures which might increase the cost of electricity claim that *linearity* cannot be proved. Linearity means that for every part of SO_2 reduced a similar proportion of Acid Rain is prevented. This cannot be proved scientifically; in no way can the SO_2 from one source be proved to be the actual gas responsible for the dead lake in Sweden – but common sense, if not linearity, tells us that the SO_2 from any chimney is having a damaging effect somewhere and it should be reduced.

One group of young people acted as 'Acid Rain Wardens' to draw attention to the contribution of car exhausts to Acid Rain.

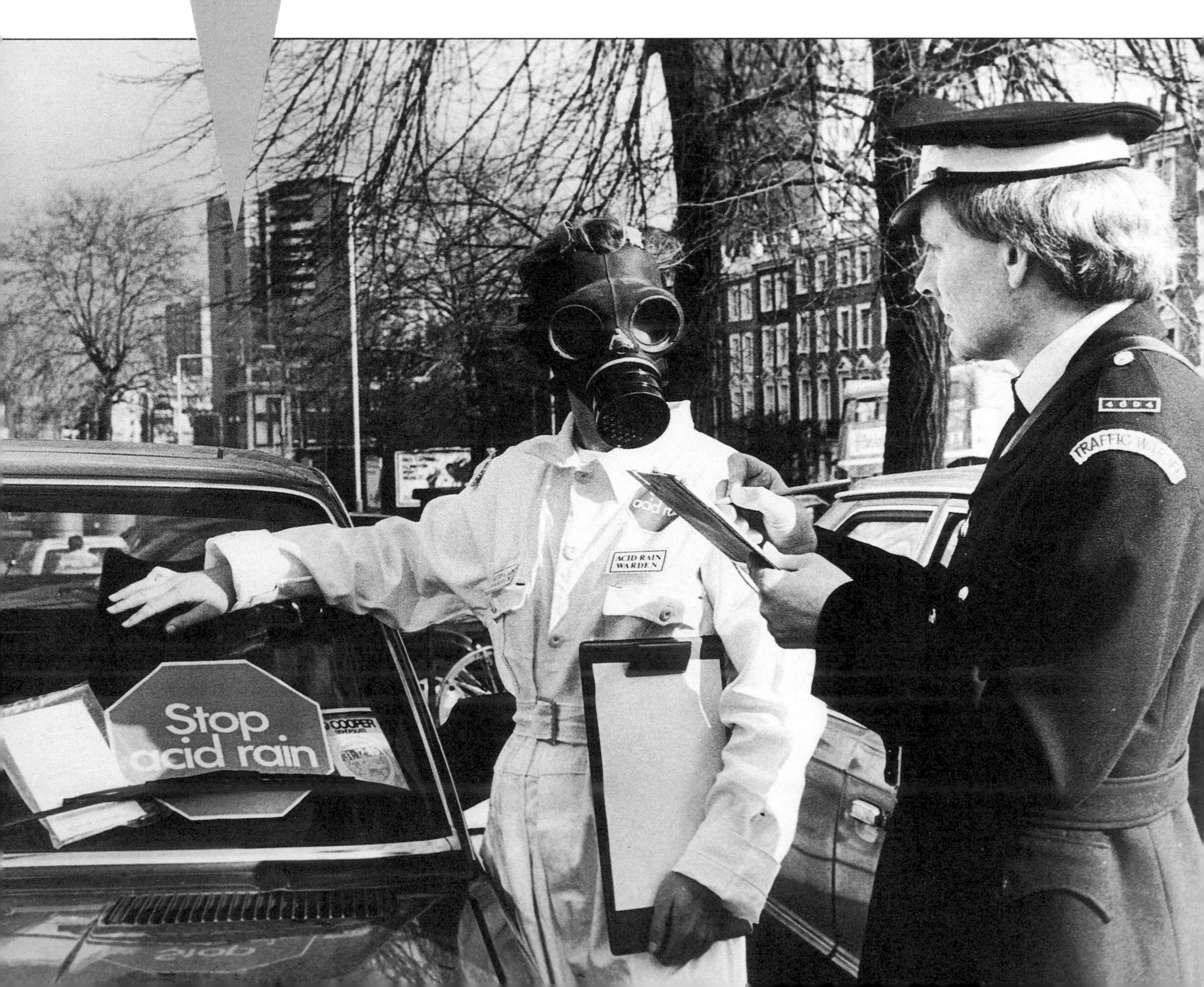

Energy conservation can not only reduce pollution, it can also save money. If for example, the electricity is not needed it does not have to be paid for and, at the same time, pollution will be reduced. It is not for me to go into great detail here about conservation; other books from Batsford have done that (*Energy, Power Sources and Electricity* is one to read). Basically it means insulation in homes, factories, shops, offices, schools and all other buildings to prevent heat escaping in winter and excessive heat entering in summer. This will result in less need for electricity to operate the heating or the air conditioning. More efficient light bulbs, switching off unnecessary lights, lowering the thermostat on the hot water tank or the radiator, using showers rather than baths, not overfilling the kettle so that every time heat is wasted on the hot water left unused – the electricity saving list is almost endless. The heat wasted when electricity is produced could be put to good use in Combined Heat and Power schemes where local homes are supplied with hot water not required by the power station: this cuts out the need to heat water in the home and so pollution is tackled twice in one operation.

The political will to reduce Acid Rain

We have heard about flue gas desulphurization: expensive if measured only in terms of money – cheap if measured in terms of a better life for people – even if they live in another country. It will be 1995 before the first FGD system in the United Kingdom is planned to be operational at the Drax power station. It is true that the price of unleaded petrol has been made less than that containing lead – a positive anti-pollution measure by the United Kingdom government – but catalytic converters were not a priority until the European Community made them compulsory for the early 1990s. Whatever the individual can achieve, governments will have the major responsibility to bring about a reduction in all forms of pollution, not only that of Acid Rain.

We must do something

Acid Rain is only one of the problems of pollution. It is a very important problem: it must be reduced. Everyone can make a contribution.

GLOSSARY AND FURTHER INFORMATION

Glossary

Acid deposition
Any dry gas or particle which falls to the ground and causes acidity or any gas or particle which has reacted with water in the air to form an acid and which falls to the ground.

Acid Rain
This term is properly restricted to acidic rainfall but is commonly used to refer to any form of acidic precipitation (snow, hail, fog, mist, rain) and even any form of acid pollution which is deposited dry. Ozone pollution is also encompassed by the term Acid Rain in many accounts of acid damage.

Acidity
A measure of the concentration of hydrogen ions.

Aerosol
Particle of water small enough to float in the air.

Air pollution
The term given to anything which contaminates the air we breathe – usually emission gases and solid particles.

Bacteria
(singular: bacterium) Kinds of microscopic single-cell organisms found almost everywhere. Some cause disease and some cause the breakdown of dead material into basic elements.

Desulphurization
The removal of sulphur, usually from fuel.

Dry deposition
Any gas or particle of matter which falls to the ground without being mixed or reacting with water.

Emission
Waste exhaust from smoke-stacks (chimneys, flues) consisting of water vapour, gases and particles.

Fossil fuel
Fuel which has been derived from vegetation or creatures which died millions of years ago. Fossils are the remains of plants and animals which died in the past and are preserved in some form or another in the rocks. Once used, fossil fuels cannot be replaced. Coal and oil are the usual fossil fuels from which we obtain coke, petrol, paraffin, coal gas, diesel oil, fuel oil and many plastic materials.

Gasoline, kerosene
American terms for petrol and paraffin.

Hydrocarbons
An important group of compounds that contain hydrogen and carbon only. There are many types and classes, including the aliphatic, aromatic, saturated, and unsaturated hydrocarbons given off through the incomplete destruction of fuel.

Nitrogen dioxide
NO_2 is a brownish gas which is formed in the atmosphere by oxidization of the colourless gas, Nitric Oxide (NO). The two together are referred to as oxides of nitrogen (NOx). NO_2 is the main chemical involved in the production of ozone.

Ozone
O_3 is a gas with a sharp odour. It is formed in the atmosphere by a complex series of photochemical (caused by sunlight) reactions where the oxides of nitrogen split into acidic atoms. Since the reaction rate depends on sunlight strength, peak amounts generally form near the middle of the day. By the time the maximum ozone concentration is reached the polluted air mass has usually moved downwind. Naturally produced ozone from the upper atmosphere can be drawn down to the lower layers and affect ground areas with a high altitude.

pH
This is the measure of acidity. Mathematically it is the reciprocal of the concentration of hydrogen ions in a substance. It follows a logarithmic scale with the lowest numbers denoting the highest acidity. Between each single pH value and the next there is a ten times difference: for example, a pH value of 4 is ten times more acidic than 5. The pH scale goes from 1 to 14, with 7 being neutral. Substances with high pH readings are said to be alkaline (or basic). The 'H' in pH stands for Hydrogen. Although the 'p' does not represent anything particular, readers may be helped to remember the idea of pH values if the 'p' is taken to mean 'proportion' – i.e. pH is the proportion of hydrogen ions present.

Photochemical smog
The term given to hazy conditions brought about by the reaction of sunlight with ozone which results in pollutant gases being trapped at ground level.

Smog
The common name given to a fog which has smoke mixed in with it.

Sulphur dioxide
SO_2 is a colourless gas with a sharp smell. Most of the man-made SO_2 comes from the burning of fuels containing sulphur and is formed by the oxidization of sulphur. Most of it is produced from the chimneys of power stations.

Temperature inversion
(or simply, Inversion) The condition which exists when there is a layer of air above the ground which is warmer than the air at ground level. Normally the air becomes cooler as you go higher.

Total acid deposition
The total amount of acidity over a given period of time. The same total could be the result of a large amount of acidity for a short spell of that period or of a small amount over the whole time.

Wet deposition
Any gas or particle of matter which falls to the ground after having mixed or reacted with water.

Useful addresses

The Acid Rain Foundation
1630 Blackhawk Hills
St Paul MN55122
USA

Acid Rain Information Centre
Department of Environment and Geography
Manchester Polytechnic
John Dalton Extension
Room E310
Chester Street
Manchester

Acid Rain Information Clearing House
Centre for Environmental Information Inc
33 S Washington Street
Rochester
NY 14608
USA

American Lung Association
28 West Adams
Detroit
Michigan 48226
USA

Department of the Environment
43 Marsham Street
London
SW1P 3PY

Friends of the Earth
26–28 Underwood Street
London
N1 7JQ

Greenpeace
30–31 Islington Green
London
N1 8XE

Information Directorate
Environment Canada
Ottawa
Ontario KIA OH3
Canada

National Association for Environmental Education
Wolverhampton Polytechnic
Walsall Campus
Gorway, Walsall
West Midlands WS1 3BD

National Society for Clean Air
136 North Street
Brighton
BN1 1RG

Schools Information Centre on the Chemical Industry
Polytechnic of North London
Holloway Road
London N7 8DB

The Stop Acid Rain Campaign/Norway
Postbox 94
N–1364 Hvalstad
Norway

The Swedish NGO Secretariat on Acid Rain
Box 245
S 401 24 Goteborg
Sweden

United States Environmental Protection Agency
215 Fremont Street
San Francisco
California 94105
USA

The Warmer Campaign
83 Mount Ephraim
Tunbridge Wells
TN4 8BS

Watch Trust for Environmental Education Ltd
The Green
Witham Park
Lincoln
LN5 7JR

World Wide Fund for Nature UK
Panda House
Weyside Park
Godalming
GU7 1XR

Resources list

Books

Acid Earth John McCormick (Earthscan Publications, 1989)

Acid Rain (Field Studies Council, 1990) (GCSE Coursework Guide)

Acid Rain – a review of the phenomenon in the EEC and Europe Environmental Resources Ltd (Graham & Trotman, 1984)

Acid Rain John McCormick (Franklin Watts, 1990) (Primary School Level)

Acid Rain John Baines (Wayland, 1989) (Primary & Lower Secondary School Level)

Acid Rain Philip Neal (Dryad Press, 1988) (Secondary School Level)

Acid Rain Controversy N Dudley, D Baldock & M Barrett (Earth Resources Ltd, 1985)

Acid Rain in the UK and Europe Steve Elsworth (Pluto Press, 1984)

The Dawton Project Peter Dawton (WWF UK, 1989)

Booklets

(Obtainable direct from organization – see Useful Addresses)

Acidification and Air Pollution (National Swedish Environmental Protection Board, 1987)

Acid Rain, The Politics of Pollution compiled by N Dudley (Acid Rain Information Group, 1983, available from FoE or Greenpeace)

Acid Rain (CEGB, 1984)

Acid Rain National Society for Clean Air – leaflet explaining acid rain

Acid Rain Schools Information Centre on the Chemical Industry – booklet for schools

Acid Rain World Wide Fund for Nature

Acid Precipitation – effects on forest and fish (the SNSF project) Norwegian Institute for Water Research 1972–1980

Autocatalysts Johnson Matthey Chemicals

Downwind – the Acid Rain story Information Directorate, Environment Canada, 1981

Lichens and air pollution (booklet and wallchart) British Museum (Natural History) or BP Educational Services, PO Box 5, Wetherby LS23 7EA

Pollution M Gittins (The National Clean Air Society)

Stop Acid Rain The Stop Acid Rain Campaign, Norway, 1986

Magazines (free to enquirers)

Acid Magazine: twice yearly from National Swedish Environmental Protection Board, Box 1302 S–171, 25 Solna, Sweden

Acid News: newsletter of the Swedish and Norwegian non-government secretariats on Acid Rain, from Swedish NGO Secretariat on Acid Rain

How many more lakes have to die?: reprint from *Canada Today* volume 12 number 2 February 1981, from Canadian Embassy

The Warmer Bulletin: monthly newsletter of the Warmer Campaign

Reports

Digest of Environmental Protection and Water Statistics Department of the Environment (HMSO)

The State of the World Environment Worldwatch Institute, 1990

INDEX